W. W. Sawyer was born in 1911. He won scholarships to
Highgate and St John's College, Cambridge, where he
specialized in quantum theory and relativity. After some
ten years spent lecturing in mathematics at various British
universities, he went to the College of Technology in
Leicester. He became head of the mathematics department,
and studied with his colleagues the application of mathematics to industry. In 1948 he became the first head of the
mathematics department in what is now the University of
Ghana.

For five years from 1951 he was at Canterbury College,
New Zealand. He founded a mathematical society for high-
school students, and this led to a significant increase in the
supply of mathematics teachers in Canterbury province.
Professor Sawyer was later invited to help in the reshaping of
mathematical education in the U.S.A. He was professor of
mathematics at Wesleyan University, Connecticut, from 1958
to 1965. At present he is professor jointly to the College of
Education and the Mathematics Department in the University of Toronto.

His books include: *Mathematician's Delight*, *Prelude to
Mathematics* (Pelicans), and *A Concrete Approach to Ab-
stract Algebra*. He is co-author of *The Math Workshop for
Children* (a textbook for junior American grades), and he
has been editor of the *Mathematics Student Journal*. *The
Search for Pattern* is the third volume in the series *Introduc-
ing Mathematics*. The first volume, *Vision in Elementary
Mathematics* (1964), and the fourth, *A Path to Modern
Mathematics* (1966), have already appeared as Pelicans; the
second volume is in preparation.

INTRODUCING MATHEMATICS
3

The Search for Pattern

W. W. SAWYER

PENGUIN BOOKS
BALTIMORE · MARYLAND

Penguin Books Ltd, Harmondsworth, Middlesex, England
Penguin Books Inc., 7110 Ambassador Road, Baltimore, Maryland 21207, U.S.A.
Penguin Books Australia Ltd, Ringwood, Victoria, Australia

First published 1970
Copyright © W. W. Sawyer, 1970

Made and printed in Great Britain
by Hazell Watson & Viney Ltd
Aylesbury, Bucks
Set in Monotype Times

Contents

Introduction

Vision in Elementary Algebra, the first volume in the series *Introducing Mathematics*, tried to present the basic ideas of algebra in a form that most adults would find easy to understand and that children could be expected to learn at a fairly early age. That volume, which from now on we shall refer to briefly as *Vision*, gave an account of how algebra arises naturally from arithmetic and described how the fundamental routines of algebraic calculation were carried out. This naturally leads to the question, 'When you have mastered these fundamental processes, what can you do with them?' This question was very much in mind when the contents of the present volume were being decided.

Those who use mathematics (or who profitably could use it, if they understood it) fall into three groups. The first group is very high-powered; it consists of mathematicians themselves and of research workers in fields such as physics and the more abstruse parts of engineering, which are so permeated by mathematics as to be in effect branches of mathematics. It seems unlikely that any of this group (except perhaps at a very tender age) will be readers of a series entitled *Introducing Mathematics*; this volume accordingly ignores applications of mathematics which are only of interest to advanced workers. A second group consists of those who are doing original, and possibly advanced work, in some field such as biology, psychology, sociology, or linguistics, and are using a limited part of mathematics in their work. A third, very numerous and important group, consists of those who are not using mathematics at all to make calculations or to solve problems, but rather as a language in which to learn some subject: they find that books suddenly introduce a formula or equation in the course of an explanation. The reader does not have to calculate by means of the formula or solve the equation, but simply to understand what message these are intended to convey.

It seemed possible that this present volume might offer some help to members of the second and third groups which they could not obtain from a standard textbook.

The material in this book was collected in the following way. I browsed through a large number of books in a public library, dealing with subjects that used some, but not too much, mathematics. Various topics were thus extracted that would serve to illustrate actual uses of mathematics. These form the skeleton of this book. To them it has of course been necessary to add some purely mathematical chapters in order to make the applications intelligible.

The arrangement of the book is by mathematical themes. Some mathematical idea is taken and then illustrated, as far as possible, by its use in various practical investigations. A reader should not be unduly disappointed if he finds that some concept, which he needs for his particular interests, is not illustrated in terms of his favourite science. Mathematics owes its existence as a useful subject to the fact that it deals with various patterns that turn up in the utmost variety of situations. The Binomial Coefficients may appear in an Indian study of poetry in 200 B.C., a seventeenth century study of gambling, or in Mendel's theory of heredity. Somebody who understands them in one of these contexts understands them in all.

One guiding principle has been that, the easier material is, the earlier it should be discussed. Thus, as far as other considerations would permit, the less use a topic made of algebraic symbolism and manipulation, the earlier the chapter discussing it would be placed. In the same way, within the chapters I have tried to treat first the simplest matters, and keep anything involving some degree of complication for discussion at the end of the chapter. Thus a reader who finds some difficulty with a point raised at the conclusion of one chapter may well find that the beginning of the next chapter causes no trouble at all.

Science is continually developing and changing. If I report that a certain investigator produced a graph in support of some contention, my comments are concentrated on the meaning of that graph and the message it seems to bear. I have not tried to make sure that no other scientist has since produced more weighty evidence pointing in exactly the opposite direction; had I done so, this book would never have been finished.

Introduction

MODERN MATHEMATICS AND MODERN SCIENCE

All the mathematical results and methods in this book are well known and are in all the traditional textbooks. In view of widespread talk about 'modern mathematics' and 'the new mathematics' this may seem to indicate a very reactionary viewpoint. However there is no necessary connexion between the age of mathematics and the age of the science to which it is applied. *Electronics in Industry* by G. M. Chute (McGraw-Hill, 1965) is surely a modern book with a modern theme, yet in it you will find a good sprinkling of familiar-looking algebraic equations. Electrical engineers still use these, even if they also use other, less familiar parts of mathematics. An equally modern title is *Transistor Circuit Analysis and Design* by J. J. Corning (Prentice Hall, 1965). Early in this book Corning says that he is not invoking higher mathematics; in his words 'the entire work is based on the premise that the reader needs background only in algebra and basic electrical circuits.' Thus, perhaps surprisingly, a considerable part of transistor design (though not all by any means) comes within the scope of this book, as being something that can be done with only a little mathematics. On this account, and also because electricity enters into everyone's daily life and should be understood, I have included Chapter 6, which illustrates the kind of mathematics involved in 'basic electrical circuits'.

The position seems to be that, while new mathematical methods may become useful in various branches of knowledge, it is extremely unlikely that the simple traditional algebra used to describe scientific laws will cease to be useful or be superseded by any branch of higher mathematics.

No topic has been included in this book simply because it is part of the traditional syllabus. In several places an attempt is made to estimate the value of some part of traditional algebra. Such discussions are necessarily incomplete, but may have some interest in view of the present widespread uncertainty as to what should and what should not be taught.

THE ARRANGEMENT OF THIS BOOK

One meets many people who profess that they cannot do mathematics. It must be rarely that this inability is due to insufficient intelligence. The most usual reasons are either a failure of communication in the teaching of the subject, or the creation of an emotional block. When there is a communication failure, the learner does not realize that mathematics is related to things already familiar to him; in consequence he is forced to regard it as a set of meaningless rules which he attempts to memorize, an undertaking in which success is extremely unlikely. It is not that there are no rules; in arithmetic, algebra, and in most other branches of mathematics, there are certain regularities, certain routines, certain procedures which can be expressed in the form of rules; the difference between the unsuccessful and the successful student of algebra is not a matter of belief or disbelief in the existence of rules. It is rather the difference between the blind following of rules and seeing these rules within some framework that makes them intelligible objects, so that, when memory falters or complications arise, the correct procedure can be thought out anew. Now it is extremely difficult to analyse and describe this framework, this background of ideas against which algebra is seen. Almost certainly it differs very much from one individual to another, and is built up to correspond to personal experience and taste. But however it is done it is essential in some way to make the work meaningful, something that can be thought about and reasoned about. The first two chapters suggest ways of doing this that may work for some. 'Mathematics through the Hand' gives some examples of how mathematics can be related to simple apparatus and physical experiences of size, weight, movement, and shape. 'Mathematics through the Eye' gives some examples of diagrams that can be used to illustrate mathematics. The first of these will appeal little to those uninterested in the physical world, the second little to those indifferent to visual perception. Chapter 3, 'Making the Curved Straight', seeks for regularities in a variety of matters, ranging from the life of the robin to athletic records; this chapter was placed early because a large part of it calls for practically no mathematical expertise, and yet arrives at a number of significant conclusions. Chapter 4 then tries to describe some of the techniques of applying algebra to a problem. Chapter 5 is purely concerned with notation; it tries to explain the meaning of certain symbolism which a reader may encounter in books on mathematics, and which, if unfamiliar, might suggest greater difficulty than really exists. Chapter 6 has already been mentioned; it uses quite simple algebra to deal with electrical and certain other situations. Chapter 7 prepares the way for Chapter 8, dealing with a procedure that has given some entertainment to various audiences and which I hope may still keep this flavour even in print. The titles of the remaining chapters speak fairly clearly for their contents. The general plan of the book is covered by two aims – on the one hand, to provide examples of mathematics (particularly algebra) being applied in actual investigations; on the other hand, to provide explanations of mathematical topics which a reader might encounter in other books, and which might hinder his progress if he was unfamiliar with them.

Most difficulties in algebra are caused by unduly rapid exposition. The author passes quickly over a step which he assumes will be obvious to the reader but which in fact is not. I have tried if anything to err in the opposite direction, to take only a few topics and to discuss them at some length; I hope this will not make the book unduly boring. This treatment of course means that the book cannot cover anything like as much ground as a formal textbook of the same size. The aim of an introductory treatment is to put a reader in the position where he can use the standard textbooks as reference works, so that he can look up for himself any topic not covered here, or find out how to deal with any troublesome detail he may run into. It is probably too much to hope that this aim will be completely achieved. Algebra unfortunately abounds in troublesome small points. In school classes you continually observe pupils asking for guidance. There may be no large principle or massive doctrine involved. It is often a little difficulty of communication; some symbol or sentence suggests to the pupil a meaning different from what the author intended, and progress is held up until this misunderstanding has been cleared up. It is almost impossible for an author to foresee what these little misunderstandings will be. This perhaps is one reason why teaching machines are not likely to replace teachers. A word from a well-informed friend is often the quickest way of resolving such a difficulty. When no such help is available, one can try reading the same topic in a great variety of books; there is a chance that one of the books will provide the needed clue.

Some of the difficulties raised by pupils in schools are due to more fundamental causes. Often pupils do not understand at all what algebra is about, so they clutch at rules and the actual appearance of the symbols. Naturally such pupils are continually in the grip of misunderstandings. Someone who has digested the ideas of *Vision* and of this book should not find himself in this situation. He should be able to cope with occasional difficulties in much the same way as a person learning a foreign language. Often in a foreign language you are not quite sure of the meaning of a particular sentence. You then consider possible meanings, and see which fits best the apparent trend of the whole paragraph. In the same way, if some step in a mathematical argument seems capable of more than one interpretation, a reader should be able to test the possible meanings, and see which of them makes sense. On page 82 of *Vision* the sentence occurs 'The most important thing in the early teaching of mathematics is that the student should form the habit of weighing evidence, of deciding for himself.' The discussion around that page, and indeed the whole book, tried to form that habit and show how it was usually possible to recognize whether a mathematical statement was correct or incorrect.

THE IMPORTANCE OF MORALE

The comparison between learning mathematics and learning a language is a fruitful one. One of the great difficulties in learning a foreign language is a sense of discouragement; there seems so much to learn and it seems so unlikely that you will ever master it. N. Potapova, in her book on the Russian language, remarks that if a beginner supplements her book by other reading material, this additional material 'should be based on the grammar and vocabulary the student knows. Otherwise new material may even retard the progress of beginner students and discourage them.' Learning a language is rather like filling a hole with plastic wood.

You first put on a thin layer. When that is thoroughly established, it becomes strong enough to support another layer, and so it continues. If you are in too much of a hurry, the whole thing breaks down. In a language, knowing one word helps you to learn others. Words occur in clusters. For instance in English, the words *guest, host, hospitality, hospitable, hospital, hostel, hostile, hostage* are all variants of the same word. The central idea is that of *stranger*. Someone learning English, and knowing two or three words of this cluster, is helped towards remembering the meanings of the others. But the cluster is no help at all to someone who is meeting all the words for the first time. In the same way in mathematics, once an idea, a process, or a symbol has become thoroughly familiar, it becomes a nucleus around which other ideas can grow. But it must be thoroughly understood if it is to play this role. Also, it must be kept in practice. Someone learning a language will naturally try to hear or to read at least a few words of that language every day. In the same way, someone who wishes to read fluently ideas expressed in mathematical symbolism needs to use some part of his mathematical vocabulary every day. It is not necessary to attack new ideas every day; using the ideas already gained until they become second nature is very much worth while. There are many ways of getting this daily practice. You may find some way of embroidering on ideas developed in this book. Chapter 3 discusses ways of spotting laws behind data; so in a different sense does Chapter 8. You may well find you can try out these techniques on figures given in the newspaper, in an encyclopedia, or in an almanac. A book of puzzles may provide exercises, as may a textbook; in either case, it is essential that the problems should not be too difficult. It is better to become fluent with something you can do than to stare for hours at something you cannot. Again, if you are learning mathematics in order to learn some other subject, you may find that articles on that subject, at a suitable level, give you the mathematical practice you need.

TERMINOLOGY AND GLOSSARY

Many of the mathematical terms used in this book will already be familiar to many readers. Rather than holding up the exposition by giving a detailed definition of every word before it was used, it seemed better to provide, at the back of the book, a list of symbols with their meanings and a brief account of the words used.

Terminology should not be taken too seriously. It is characteristic of poor teachers that they lay great stress on words and little stress on ideas. They do this because they teach by rules, and rules are expressed in words. Effective thinking in mathematics depends not on remembering words but on understanding situations. A good teacher does not need a large vocabulary of technical terms; he can point to the board and speak of 'this number' and 'that symbol'. A book unfortunately cannot point, and is therefore forced to use a certain amount of jargon in order to specify which number or which symbol the reader should direct his attention to.

CHAPTER ONE

Mathematics through the Hand

As was mentioned in the Introduction, one way of making mathematical ideas vivid is to associate them with our experiences of the physical world. This chapter describes a number of pieces of apparatus and simple experiments that can be used for this purpose. Most of the devices here described were developed at Leicester College of Technology in the 1940s. A whole book could be devoted to describing how a complete mathematics course may be linked to scientific experiments and demonstration apparatus. The present chapter can only give, in a rather unsystematic way, some samples of what can be done. This approach is now being widely used in England with younger children. A useful account of it can be found in the booklet *Mathematics in Primary Schools* published by the Ministry of Education and Science. The title is somewhat misleading; this booklet would be of interest, for instance, in many secondary schools.

SOME APPARATUS

A scientific experiment, to be useful for a mathematics lesson, must use apparatus that is simple, cheap, and reliable. It should be cheap, so the teacher is not worrying all the time about it getting broken. It should be reliable, so that the time is spent on discovering the central principle and not on correcting a variety of mechanical faults.

At Leicester we often used to introduce engineering apprentices to algebra through a device known as the Platform on Rollers. The rollers were a pair of cotton reels, held together by a frame to which a pointer was attached. The platform was simply a flat piece of wood, resting on top of the cotton reels. It too carried a pointer. The reels rolled on another flat piece of wood, part of which was marked out in inches. To begin with, the two pointers were both at the 0 mark, as at (a) in Figure 1.

Anyone who has ever moved a heavy object by placing it on

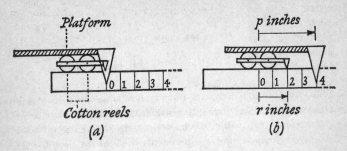

Figure 1

rollers knows that you continually have to take the rollers which emerge behind the object and keep putting them in front. In fact, the object continually overtakes the rollers; it goes faster than the rollers. How much faster? As a rule, nobody knows. However, one can quickly find out, by observing what happens when the platform is pushed forward, as in (b). About a minute is usually sufficient both to make the experiment and to spot the law. It will be found that by the time the rollers have gone 1 inch, the platform has gone 2 inches. When the rollers have gone 2 inches, the platform has gone 4. When the rollers have gone 3 inches, the platform had gone 6. It does not take long to notice that the platform always advances just twice as far as the rollers.

Now there is no doubt that a student should record any experiment he performs. This not only puts the experience in permanent form; it also forces the student to think just what he did and what conclusion it led to. However, writing out a long account can be very wearisome, and it is also ineffective, since, on any future occasion when one has to refer to it, it takes a long time to read. The best notes are brief and present the essential points so thay can be taken in at a glance.

It was our custom at Leicester after the class had discovered the law, to write on the blackboard, 'The number of inches the platform goes is twice the number of inches the rollers go'. We would then say, 'This is rather a long statement. Can you suggest any abbreviations?' The apprentices would make one suggestion after another; the statement would be written again and again in shorter and shorter form, and eventually would have been condensed to $p = 2r$, the letter p being all that survived of the phrase 'the number of inches the platform goes' and r of 'the number of inches the rollers go'. The students would then draw in their notebooks a diagram somewhat resembling Figure 1 (b), and write next to it the observed law, $p = 2r$. This was their first contact with algebra, and they met it as a useful device for saving a lot of writing and for recording something they already understood. Incidentally, in a space of a few minutes, they were retracing an evolution of algebra that originally occupied centuries. The first mathematical statements were in full literary form, as in 'The square of the hypotenuse is the sum of the squares on the other two sides.' The transition from this form, known as 'rhetorical algebra' through 'syncopated algebra' to the modern form of 'symbolic algebra', $a^2 + b^2 = c^2$, was a long and painful process.

If algebra is approached by a series of experiences similar to the one just described, several ideas will be conveyed without ever being mentioned in words. For instance, many beginners in algebra find difficulty with the idea of a variable. They may be quite happy with π, which stands for a fixed number that is somewhere around 22/7 or 3·1416. They feel that x and y also should be code names for fixed numbers, and are disturbed when $x = 3$ in Question 1 but $x = 10$ in Question 2. Someone who meets algebra through miniature scientific experiments in which things actually are varying is much less likely to experience this difficulty. If r is an abbreviation for 'the number of inches the rollers have gone', it is clear that r may be 0 or 1 or 2 or 3 or values between these. We arrive at the law $p = 2r$ only by examining the *whole of the table*.

| Distance gone by rollers (inches) | 0 | 1 | 2 | 3 |
| Distance gone by platform (inches) | 0 | 2 | 4 | 6 |

If we were confined to one situation, for example that frozen in Figure 1 (b) which might convey the mistaken impression that we were only concerned with $r = 2$, $p = 4$, we would be unable to make any guess about a general law. The single case $r = 2$, $p = 4$ is consistent with $p = r + 2$ or $p = r^2$ just as well as with $p = 2r$.

A mathematician might object that there are other laws that fit the whole of the table above. After a little work, he might arrive at the suggestion $p = r^4 - 6r^3 + 11r^2 - 4r$, which, as you can check, does in fact give 0, 2, 4, 6 for p as r runs through 0, 1, 2, 3. If the mathematician is making a purely logical point, he is perfectly correct. We have in no sense *proved* that $p = 2r$ is the only law that describes the movement of the platform and rollers. But for the teaching of algebra and science, this objection should be largely ignored. Science never proves anything; it makes guesses, and goes by them so long as they work well. Certainly no beginner of algebra is going to produce as complicated a formula as the one above. To a normal human being, the table above suggests very strongly that we double r to get p. We are using algebra as a convenient way of recording this *guess*. And this is a very effective way of introducing people to algebra and showing how it is used and giving them a feeling for the meaning of a formula. The guessing game and guessing machine in Chapter 4 of *Vision* used essentially the same approach, but without the relation to practical experiments.

The platform and rollers experiment also lays foundations for the idea of 'function'. It is clear to anyone who handles the apparatus that, if you choose a position for the rollers, you have fixed the position of the platform. I am assuming that the apparatus is used as intended; the platform is not allowed to slip over the rollers, nor is it taken off and put back in a different position. This appears in the arithmetic. If someone tells you that r is 5, you know that p must be 10. In a situation of this kind, mathematicians use the word 'function'. If there are two things, x and y, which are connected in such a way that choosing x fixes y, then we write $y = f(x)$. Here f is the first letter of the word 'function', and the symbols imply that there is some procedure or operation, f, that applied to x will give y. For instance, with the platform and rollers, if we are given r, we double it to get p. The operation here is that of *doubling*, which is a very simple one, easily expressed by a formula. But, as it is used today, writing $y = f(x)$ does not imply that there is any simple calculation that gives y. The operation f may involve using a graph or a table, or feeding the specification of x into an electronic computer and getting a specification of y. All that $y = f(x)$ means is that there is some law giving y when x is known. You tell me x, I will tell you y. How I do it is my business; there is no guarantee that I am using a simple formula.

The mathematical idea of function is closely related to the scientific idea of law. There is no point in trying to find a law connecting x and y if in fact they are not connected. For instance, someone might consider the question, 'What formula gives a person's weight in terms of his height?' There is no answer to this question. A man five feet high is rather shorter than the average, but I have no way of telling whether he is a short thickset man or a short thin man. You could have many men, all five feet high, and with widely different weights. A scientist might say, 'There is no law relating weight to height.' A mathematician might say, 'There is no function that gives weight when height is known.'

When we speak of a scientific law we usually imply something with fairly wide applications. I would hardly speak of a law for the amount I spent on bus fares on the various days of the past week. But I could properly speak of a function. Providing I have kept records, if you ask me what I spent last Tuesday, I can tell you. To each day, there corresponds a definite amount spent in buses on that day.

The last example of a function meets the formal, logical definition of what a function is, but it is certainly not an example of a function with mathematical significance or scientific importance. Learners of mathematics may well become confused if time is spent discussing such examples; they may feel there is some deep principle involved which they are missing, and be puzzled. They are not missing anything; there is nothing to miss. Some mathematicians and teachers have laid great stress on precise verbal definitions of 'variable' and 'function'. This, I believe, is a great mistake. Most people do not learn from verbal definitions. They learn by actually using something, and the meaning gradually sinks into them without the intervention of words. If a beginner at algebra performs a series of experiments, in which things are varying and he is seeking for a law connecting these things, he will unconsciously come to appreciate that p and r are variables, and that there is a law, a function, indeed in this case a formula, relating p to r.

LINEAR FUNCTIONS

The equation $p = 2r$ defines one of the simplest functions there is. There are quite a number of simple experiments that can be performed successfully by anyone capable of recognizing the operation of doubling.

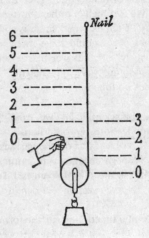

Figure 2

Figure 2 shows a pulley being raised by a hand. The experiment, which should only take a very short time, has the object of finding how the height the hand rises is related to the height the pulley rises. The result will be recorded, as before, in algebraic shorthand. This experiment can be demonstrated to a class. The horizontal lines are chalked on a blackboard, about 4 inches apart. The chalk marks should be fairly thick; we do not want someone saying, 'I think your hand has risen three and one-sixteenth divisions'.

Figure 3 shows a simple piece of equipment which not merely allows one to discover a simple law; it also establishes a bridgehead on the subject of trigonometry. First of all a thin strip of wood or cardboard is fastened to a board (such as a drawing

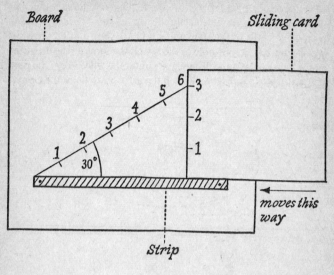

Figure 3

board). A line is then drawn on the board, at an angle of 30° to the strip, and marked out in inches. In the illustration, this line carries the figures from 0 to 6. A card is then placed on the board and slides with its base in contact with the strip. Along a vertical edge of the card inches are marked out, as shown in the figure by the numerals 1, 2, 3. In the position shown, the 3 on the card is in contact with the 6 on the oblique line. If the card is pushed to the left, there will be a stage at which the 2 on the card meets a figure on the oblique line, and later still the 1 on the card will meet another figure on the oblique line. The law relating the figure on the card to the figure it meets on the oblique line is sought.

It will be seen that the upper edge of the strip, the left edge of the card, and the oblique line always form a right-angled triangle. By slipping the card along, the scale of this triangle is altered, and we observe the effects on the lengths of the oblique and the vertical sides.

Another easily made device is shown in Figure 4(a). A spike (such as a nail) sticks up in the middle of a flat board. A thin strip of wood or metal has holes bored in it at equal intervals. A small wheel is attached to the end of the strip, as shown in the illustration. The details of the strip and wheel are shown in Figure 4(b). The distance from one hole to the next is the same as the diameter of the wheel. The same distance separates the wheel from the hole nearest to it. So in Figure 4(b) all the arrows are the same length.

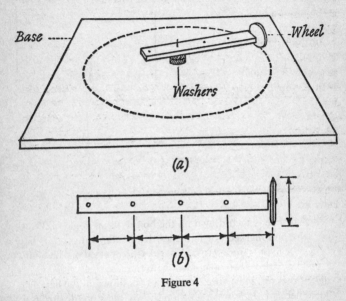

Figure 4

Some washers are placed around the spike so that, when the apparatus is assembled as in Figure 4(a), the strip will be level. The strip is now free to rotate about the spike, so the wheel will follow a circular track, shown dotted in the diagram. A mark is made on the wheel, and we observe how many times the wheel turns as it goes once round the dotted track. If there is no slipping, this should turn out to be a whole number. The spike is placed through each hole in turn, the strip is rotated, and the number of turns made by the wheel is observed and recorded. A law is then sought connecting the number of turns with the number specifying the hole used. The hole nearest the wheel is hole number 1, the next is number 2 and so on. If we take the distance between one hole and the next as our unit, the number of the hole gives the radius of the track described by the wheel. In Figure 4(a), the spike is in hole number 3, and the radius of the dotted circle is 3 units.

The wheel is in effect being used as a measuring wheel. A teacher could introduce this experiment by some discussion of why athletes try to get on the inner track when racing in a circular arena, or why a differential has to be provided in the rear axle of a motor car.

All the experiments so far have led to results of the type $y = 2x$. Naturally, there are other experiments which lead to results such as $y = 3x$ or $y = 4x$, or, more likely if the experiment has not been carefully rigged to give a simple result, something like $y = 3 \cdot 29x$. All such results can be lumped together in the more general form $y = ax$, where a is some fixed number depending on the nature of the experiment. There are other experiments which lead to equations like $y = 2x + 5$ or $y = 2 \cdot 4x + 1 \cdot 7$. These belong to the family of equations $y = mx + c$ which was discussed on pages 307-9 of *Vision*. There we saw that every such equation had a straight line for its graph. The equations accordingly are called *linear* equations and the resulting relationship between x and y is called a linear function.

THE STAIRCASE EFFECT

A physical illustration of a linear function is shown in Figure 5. We have several identical jars, of cylindrical shape, and a small can. Into the first jar we pour one can of water, into the second

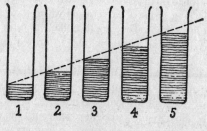

Figure 5

jar two cans, and so on. When the jars are placed at equal intervals, we see something that resembles a staircase. All the steps are the same height. We rise the same amount whether we go from the first to the second jar or from the fourth to the fifth. Incidentally, it does not matter if, as often happens, there is some irregularity in the bottom of the jar. So long as the sides of the jar are uniform, each can added will raise the water level by the same amount. This is the characteristic feature of a linear graph.

We have been studying a process that occurs by jumps. At each stage, a complete can of water is poured in. Such jumps are usually easier to imagine than a process that takes place continuously. But Figure 5 could also illustrate a continuous process. Suppose a tap is turned a little, so that water runs slowly but steadily into a jar; we can then think of Figure 5 as representing not five different jars, but the same jar at five different times. The drawings could be pictures of the jar taken when the water had been running for 1, 2, 3, 4, and 5 minutes respectively. The dotted line is intended to indicate that the process of filling the jar continues steadily between the times when we observe the state of the jar. This dotted line is in effect a graph from which we could read off the height of the water at any time. Thus the staircase graph, appropriate to a discontinuous process changing by jerks, helps us to visualize the process of continuous, steady growth. The graph of the latter rests on the graph of the former like a plank lying on a staircase and touching each step.

LAWS WHICH ARE NOT LINEAR

As was mentioned on page 279 of *Vision* it is dangerous to subject learners to a long series of exercises in which they plot nothing but straight line graphs. They are then likely to form the habit of plotting two points and using a ruler to draw the line joining these.

One way to meet this is to suggest that, as an insurance against errors of calculation, three points should always be plotted before the ruler is used. If the law happens not to be linear, it will, as a rule, be impossible to put a straight edge against the three points.

In any case, we should make sure that learners meet examples of functions and graphs which are not linear. One simple example

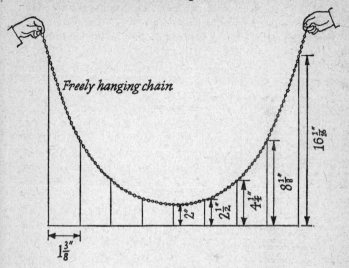

Freely hanging chain

$16\frac{1}{16}$"

$8\frac{1}{8}$"

$4\frac{1}{4}$"

$2\frac{1}{2}$"

2"

$1\frac{3}{8}$"

Figure 6

is the graph of a hanging chain, illustrated in Figure 6. The lengths of the upright lines are obtained as follows. In the table below, each number in the top row is double the number before it. The figure 1 occurs in the middle of the table. The second row is simply the first written backwards. The third row is obtained by adding the first two rows. It gives the number of inches in the upright lines.

$\frac{1}{16}$	$\frac{1}{8}$	$\frac{1}{4}$	$\frac{1}{2}$	1	2	4	8	16
16	8	4	2	1	$\frac{1}{2}$	$\frac{1}{4}$	$\frac{1}{8}$	$\frac{1}{16}$
$16\frac{1}{16}$	$8\frac{1}{8}$	$4\frac{1}{4}$	$2\frac{1}{2}$	2	$2\frac{1}{2}$	$4\frac{1}{4}$	$8\frac{1}{8}$	$16\frac{1}{16}$

The upright lines should be spaced a distance of $1\frac{3}{8}$ inches apart.*
If the graph paper is then fastened to the wall, it should be possible to hold a chain so that it hangs freely and passes just in front of the points at the tops of the upright lines.

*Strictly speaking, the number should be 2 log$_e$ 2 which is 1·386. The approximation given involves an error of just over one-hundredth of an inch.

Figure 7 shows another simple piece of apparatus with a non-linear law. It is a primitive water clock made from two tin cans. One can is slightly narrower than the other, and slides inside it in

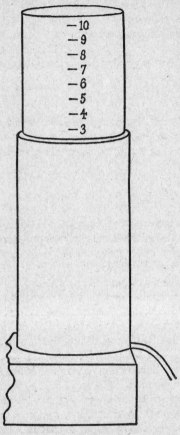

Figure 7

the manner of a piston. The upper can floats on water in the lower can, but descends gradually since the water is allowed to escape through a hole near the bottom of the lower can. Evenly spaced marks are made on the floating can, and the time is noted at which each of these disappears behind the rim of the lower can. A graph is made, connecting the number attached to each mark with the time at which that mark disappears. This graph will be curved, since the water runs out faster at the beginning of the process than at the end.

There may be difficulty in finding two cans with only slightly differing diameters. If there is a big difference, the floating can tends to tilt. This can be overcome by binding some adhesive material around the lower part of the floating can. Obviously we do not wish to obtain an airtight fit which would hinder the movement. We also wish to keep the floating part as light as possible; if it is heavy, its weight exerts a considerable constant pressure on the water, and this masks the effect of the changing head of water. As a result, the graph may become only slightly curved.

The curve in which the water spurts out from the hole is also of interest. Figure 8 is made with the help of graph paper marked in quarter-inch squares. Since the water is falling, our vertical measurements are downwards. Lines are marked at depths 1, 4, 9, and 16 spaces below the starting point 0. To draw curve A we move half an inch across the paper each time we plot a point. To draw curve B, we move 1 inch to the right at each step, and for curve C we move $1\frac{1}{2}$ inches across each time. All the points are plotted on the horizontal lines already marked. If this graph paper is held just behind the jet of water emerging from the can, it will be found that there is a time when the jet just hides curve C; later it hides B and last of all curve A. If the floating can is removed, so that water is flowing out of the fixed can without any extra pressure, I find curve A can occur when there is less than 1 inch of water remaining in the can.

It might be possible to demonstrate that the jet really has the shape stated by putting various barriers in the way of the jet, but each barrier having a hole so placed that at a certain stage the water would pass through all the holes and emerge on the other side of the barriers.

In Figure 8, if we measure x across and y down, with the spaces of the graph paper as our units, graph A will have the equation $4y = x^2$, while B has $16y = x^2$ and C has $36y = x^2$. These equations are helpful if you want to plot some points between

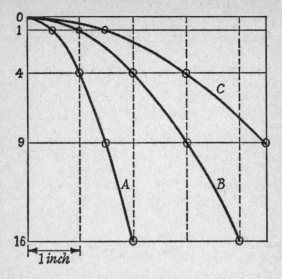

Figure 8

those shown in Figure 8 in order to draw the curves more accurately.

Figure 9 shows an equally inexpensive piece of apparatus, to study how a rubber band stretches when increasing weights are hung from it. If regular weights are available, they can be placed in the can and the corresponding depth to which the can sinks observed. The ruler can be placed as shown, or, if preferred, the length of the rubber band can be measured directly. If weights are not available, water can be put in the can. Any small container is used; it is filled with water, which is poured into the can; this is done many times. After each pouring, the length of the rubber band is observed. It will be found that at first the length increases by equal steps (linear law) but that as the weight in the can increases, the rubber starts to give way and the steps increase in length. It is instructive to draw a graph and observe the initial straight part, and also the way in which the graph later departs from this straight line.

It may be found that measurements are easier to make if,

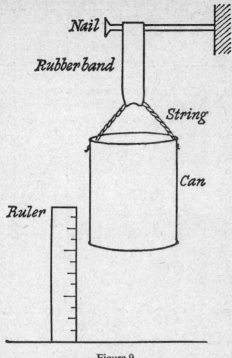

Figure 9

instead of using a single rubber band, a long spring is made by tying several rubber bands end to end.

CAR STEERING

Some years ago I was told of a primary school teacher in England who felt it would be good if her children learnt some French. An inspector happened to visit her room as she was giving her first lesson. She was on her bicycle and was cycling around the room; the children were standing up like traffic policemen and as she approached would call out, '*À droite!*' or '*À gauche!*' and she would turn right or left accordingly. It would, I imagine, be rather exhausting to design and conduct an entire French course on these lines, but as an initial lesson this was wonderful. One can imagine the children going home and telling their families 'If we said *à droite* she went right and if we said *à gauche* she went left.' The lesson would create the impression that learning French and probably all the other lessons with this teacher were going to be interesting, and this feeling would persist even when some fairly routine piece of work was being done. The first lesson in any subject tends to fix the emotional colour of that subject, and it should be designed at least as much for its dramatic effect as for its intellectual content.

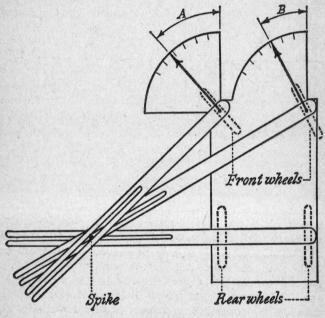

Figure 10

The piece of apparatus about to be described is based on the idea that many boys are interested in cars, and that drawing a graph to establish a certain fact of importance in car design might produce both attention and favourable associations in their minds. Figure 10 shows the device. A flat piece of metal represents the body of the car. The wheels are underneath this piece, so they can roll on the table. A spike, sticking up from the table, represents the point about which the car is turning. There are also various strips, the role of which will be explained in a moment.

Earlier, in Figure 4, we had a wheel connected to a spike by a strip so that the wheel would run round a circle. Naturally, the wheel was fixed at right angles to the strip, as in Figure 11(a). If it had been fixed as in Figure 11(b), it would not be running

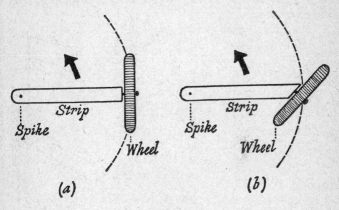

Figure 11

smoothly but would be slipping sideways all the time. Obviously in a car or other vehicle, we wish the wheels to run as far as possible without sideslip. The strips in Figure 10 are arranged so that, in the model, the wheels take the ideal positions that are theoretically desirable. One strip represents a continuation of the rear axle. This ensures that, as the whole device rotates about the spike, the rear wheels do not slip. Each of the front wheels has its own strip; being fixed to this strip at right angles ensures that the wheel rolls without sideslip. A pointer is fastened to each strip, to indicate the position of the wheel, and protractors are provided, so that the angles through which the wheels have turned can be observed and recorded. These angles are marked A and B. It is clear from the figure that, except when both angles are 0°, the angles A and B are not equal. To investigate the relationship

31

between these angles, the spike is placed at various distances along the extension of the rear axle, and the resulting angles A and B are noted. A graph is then drawn to show the various points (A, B).

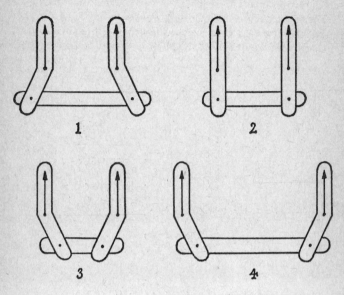

Figure 12

This work was used as an exercise in the National Certificate course at Leicester to bring out the meaning and use of a graph. The idea probably could be carried much further. In actual practice, of course, one cannot have huge strips projecting from the side of a car and interfering with other traffic. We want to produce the same effect by a linkage that will fit neatly inside the car. One could investigate this by taking a variety of linkages of the types indicated in Figure 12 and trying to discover which type and what proportions gave a graph closest to the ideal graph found from the model. It is evident that No. 2 is not the ideal linkage, since it would make B always equal to A. Numbers 3 and 4 belong to the same type; they are shown to emphasize that it is not only the type, but also the proportions between the various members, that determines how B will depend on A.

BINOMIAL COEFFICIENTS

The apparatus so far discussed has related to mechanisms or to physics. Figure 13 shows a somewhat different type of device, not related to any physical application, but simply intended to show a

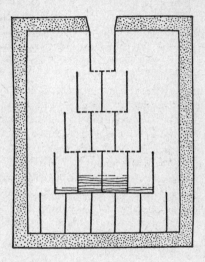

Figure 13

certain collection of numbers in a vivid manner. This device is due to W. H. Joint, who initiated the laboratory approach to mathematics at Leicester College of Technology. The numbers it illustrates are those in what is often called Pascal's Triangle, namely

$$
\begin{array}{ccccccccc}
 & & & & 1 & & & & \\
 & & & 1 & & 1 & & & \\
 & & 1 & & 2 & & 1 & & \\
 & 1 & & 3 & & 3 & & 1 & \\
1 & & 4 & & 6 & & 4 & & 1 \\
\end{array}
$$

We shall be discussing these numbers in the chapter on the Binomial Theorem, and mentioning their relevance to problems of heredity and of statistics.

The apparatus has a pane of glass at the front. Behind the glass can be seen various metal containers, arranged as shown in Figure 13. The floors of these containers are removable. They can be pulled out from behind the cabinet. In the diagram, the dotted lines indicate floors which have already been removed. Initially, all floors would be in position, and the top compartment would be filled with some suitable material, such as sand, small beads, or dried peas. When the floor of this compartment is pulled out, the contents fall into the two compartments below. Each compartment receives approximately the same amount, and at this stage we have the material divided in the ratio 1 to 1. The floor of the two compartments, where the material now is, consists of a single piece of metal. When this piece is pulled out, the material falls into the three compartments at the next level. The middle compartment receives contributions from both the compartments at the second level, while the end compartments are fed from only one. Thus at this level the material divides itself in the ratio 1:2:1, and illustrates the third row of the triangle of numbers. These three compartments again have a single sheet of metal for their floor – it is not possible to release the material from one of them without at the same time releasing it from the others. When this floor is removed, the material falls into the four compartments at the fourth level, and we have reached the situation shown in the illustration. It will be found that the material is now shared out in the proportions 1:3:3:1. At the next stage of the demonstration, as you would expect, the remaining floor piece is removed and the material divides itself among the five compartments of the fifth level in the proportions 1:4:6:4:1, as in the fifth row of the Pascal triangle. There is of course no need to stop at this point; the apparatus could be extended to show more stages if this were desired.

The advantage of this device was that it gave a class something to look at and think about during a discussion of the Binomial Theorem, and conveyed a certain familiarity with and feeling for the corresponding distributions of material, which simply writing the numbers might well have failed to give.

THE ALGEBRAIC BALANCE

The device shown in Figure 14 was used right at the beginning of an algebra course. The idea was due to R. Kitchen. He argued, 'The books always say that an equation may be likened to a balance. But we know that to many students words mean very little. Why talk about a balance? Why not have a balance there in the room?' Accordingly he constructed a balance, on which a simple equation could be represented. For weights Whitworth nuts were used. Matchboxes were marked x, and a teacher would

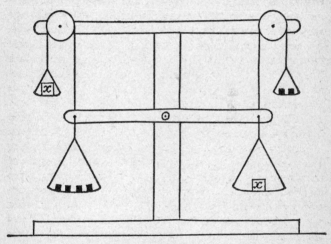

Figure 14

put nuts into these boxes, the same number in each. Even a home-made balance is remarkably sensitive, and this sensitivity had to be cut down by putting in some extra friction; otherwise the scale might appear to be out of balance simply because one nut was minutely heavier than another. Pulleys were provided above the beam of the balance, so that 'negative weights' pulling up-wards could be arranged. The situation shown in Figure 14 corresponds to the equation $4 - x = x - 2$. On the left, we see four nuts pulling down and a matchbox x counteracting them; on the right, a matchbox x pulls down, with two nuts counteracting it. The best way of using the balance is for the student to write down what he thinks the next step should be, and then to alter the weights on the balance so as to represent the new equation. If he has made a mistake, the scales will immediately swing over and indicate that the equation no longer balances. Thus the student is immediately notified that he has taken a false step, and so he learns quickly from the actual experience. If the student worked correctly with the equation $4 - x = x - 2$ he might reach in turn $4 = 2x - 2$, $6 = 2x$ and finally $3 = x$. At this stage the matchbox would be opened and found indeed to contain three nuts.

CHAPTER TWO

Mathematics through the Eye

IN some schools, graphs are treated as a kind of museum piece. Rather late in the syllabus, there is a chapter on graphs; the idea seems to be that in some examination you might get a question about the graph of an equation. This is entirely the wrong way to approach graphs. As was mentioned in the introduction, success in mathematics is very much a matter of having a feeling for what is going on, not blindly following the steps of the formal process but seeing the work in the framework of some intelligible situation. The first thing to do in approaching any piece of mathematics is to cast around for some way of appreciating what is involved, some way of making the question vivid to yourself so that you will be able to think about it just as you think about familiar things in everyday life. Graphs are one way of bringing out the meaning of a mathematical situation. They are not a magic key to all mathematics, but they are one of the things that should always be considered as a possible approach. Graphs therefore should permeate all mathematical work; one should form the habit of perpetually asking oneself. 'Would a graph help to make this more meaningful?' The more you do this, the more helpful graphs become, since you gradually build up a sort of album of familiar graphs; when you have met the same graph in half a dozen situations, it naturally conveys more to you than it possibly can the first time you meet it. It has acquired all kinds of extra associations; it has become familiar. It is important to realize that this familiarity, the feeling that you understand the graph and know what it is all about, is as much an emotional as an intellectual thing. It comes only from repeated use. It is no use, the first time you see a graph, making an immense effort to dig out its secret message. This will only discourage you. I cannot find anything to put into this chapter that strikes me as at all unusual or profound. But I believe that anyone who repeatedly uses the simple ideas in this chapter will eventually gain very much in confidence and insight.

The Search for Pattern

What then can we learn by looking at a graph? Among the first things to look for are high points and low points – mountain peaks and valley bottoms. Figure 15 is taken from page 16 of C. O. Carter's *Human Heredity* (Penguin Books, 1962). It shows how the likelihood of giving birth to twins is related to a woman's age. The first thing that strikes us is the peak around 37 years of age. A woman who becomes pregnant between 35 and 39 is more

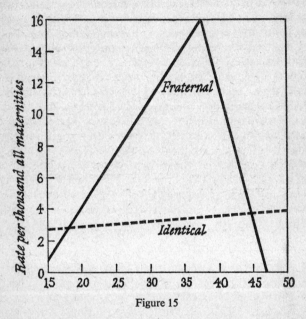

Figure 15

likely to give birth to twins than either a younger or an older woman. I believe that graphs have become so much a part of our civilization that this will be evident to most readers, and does not call for explanation. Yet we are certainly not born with the ability to interpret such a graph; there must be a time when one first sees the meaning of such a diagram. This comes presumably from the experience of making graphs for oneself of things which are interesting and familiar, and in schools quite young children are increasingly being given the opportunity to do this. If some-one questioned our interpretation of the twinning graph, we could

I suppose, point to the figures; at 37 years there are 16 chances in 1,000 of a pregnancy leading to twins, as against 2 somewhere in the teens or around the age of 45.

In this same figure, the dotted line shows the chance of producing identical twins. This line does not have a peak. The chance appears to rise steadily, though very slowly, with increasing age. In fact the change is so slow that Carter simply speaks of the rate as being 'fairly constant – between three and four births in a thousand'. It may be that the apparent rise is not significant; one would have to go back to the original data to judge this.

It will be noticed that we have passed quite naturally from the topic of peaks and valleys to the question whether something is rising or falling, and how rapidly. For fraternal twins the graph climbs the hill between the ages of 15 and 37 and descends after 37. The descent is steeper than the climb. This means that the likelihood of twins decreases more rapidly after 37 than it increases before that date, and indeed the figures bear this out; to rise from near 0 to 16 takes 22 years, but the descent from 16 to 0 is accomplished in about 10 years. The steepness or gradient of the graph tells us how fast the quantity in question is growing – or shrinking, if the graph happens to be falling. This remark can be illustrated by Figure 16, taken from G. E. Fogg, *The Growth of*

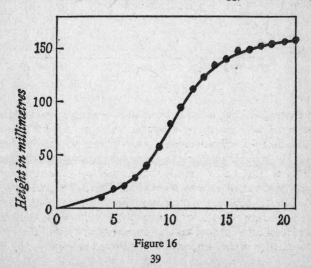

Figure 16

Plants (Penguin Books, 1963). It shows the growth of a lupin seedling. For the first five or six days the growth is at a moderate rate. Then the graph becomes steeper, indicating that the plant is growing faster. The maximum steepness occurs around the tenth day. Later the speed of growth slackens, as shown by the graph becoming less steep, and finally the graph becomes flat; the plant has reached its mature height; growth in height has ceased.

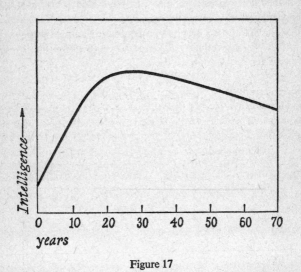

Figure 17

Figures 17, 18, and 19 are taken from Cattell's *The Scientific Analysis of Personality*. They purport to show how a person's intelligence, social confidence, and enthusiasm are likely to change in the course of a lifetime. It will be seen that intelligence seems to reach the top of the mountain around 25 years of age, and to go steadily downhill until one becomes old enough to be a Prime Minister. Social confidence tends to rise steadily; as you grow older you become less shy, more thick-skinned. Enthusiasm, on the other hand, seems to sink steadily after the age of 10. It should be borne in mind, as Cattell himself points out, that these curves are based on averages. They show what happens in most cases as the individual receives the normal amount of battering from life.

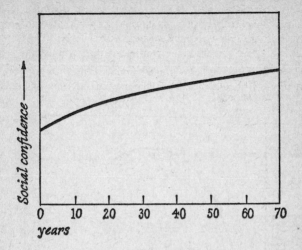

Figure 18

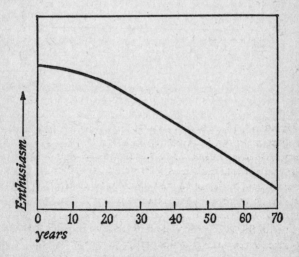

Figure 19

They need not apply to those who are particularly sheltered, particularly successful, or particularly tough. Those whom the gods love may still be young at 70.

Cattell also gives a graph indicating the kind of relationship that may be expected between the amount of anxiety a pupil feels and how well he does at school. This graph is shown in Figure 20.

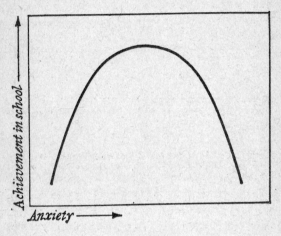

Figure 20

When anxiety is very low, achievement is also very low. This corresponds to the pupil who does not care at all whether he succeeds or fails. A pupil with moderate anxiety does better; he is worried by the thought that he may not do well, so he works and pays attention. But if anxiety passes a certain stage, achievement begins to fall again; the pupil does badly because he is worrying too much. Very high anxiety can produce a kind of paralysis and the pupil achieves almost nothing.

Figure 21 is taken from *Stability and Change in Human Characteristics* (Wiley, 1964) by B. S. Bloom, of the University of Chicago. The graphs show the I.Q. scores of Negro pupils, some of whom have grown up in Philadelphia, while others have come to Philadelphia at various ages from the South. Graph (a) is for children who have grown up in Philadelphia. The wobbles in the

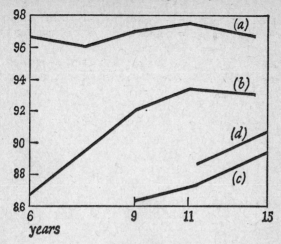

Figure 21
(adapted from B. S. Bloom, *Stability and Change in Human Characteristics,* Wiley, 1964)

graph are presumably not significant. Essentially the graph is level, corresponding to steady scores around 97. Graph (b) is for children who came to Philadelphia at 6 years of age. For them the average I.Q. climbs steadily from 87 to 93 and then apparently fails to climb further. Graphs (c) and (d), for children coming at ages 9 and 11 respectively, both show an upward slope. The interpretation of I.Q. scores is always a tricky business. An I.Q. test may determine simply how far a child is aware of the things that are recognized by the child of a prosperous family in a big city. They may not test at all a child's grasp of Negro culture or his experience of the rural South. On the other hand it is possible that Philadelphia, while being far from providing an educational paradise, may awaken mental powers that would remain for ever dormant in the South. The truth may well contain elements of both these contentions.

Figure 22 is taken from Carter's book, *Human Heredity,* that has already been quoted. It also deals with the effect of a changed environment on I.Q. scores; it is concerned with adopted children. The question investigated is this; suppose there is a group of

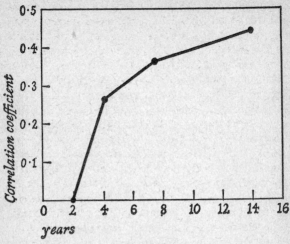

Figure 22

adopted children and you want to predict which of them will be the cleverest. Which will it be more helpful to know – the I.Q.s of the children's actual mothers, or the educational level of the adopting mothers, the women who are actually looking after them? The only difficulty in understanding this graph is due to the rather abstract nature of the quantity it portrays. It is *not* a graph showing how the children's I.Q. scores increased. It is a graph of a correlation coefficient, something that will be explained in a later chapter of this book. This coefficient is a number that tells you how closely two things are connected. The two things that are involved here are the achievements of the children on the I.Q. test and the achievements of their real mothers. A correlation of $+1.0$ would mean that the child with the highest score had the mother with the highest score, the child with the second highest score had the mother with the second highest score, and so on, right down to the child with the lowest score having the mother with the lowest score. Such a result is extremely unlikely, for it would mean that the father's intelligence had nothing at all to do with the child's. The largest number we should expect to find for the relation of children's performances to mothers' is 0.5, and in fact in normal homes (where there has been no adoption) the correlation between the intelligence of one parent and that of a child is around 0·5. A correlation of 0 would mean there was no connexion. A correlation between 0 and −1 would indicate an opposite connexion. Such a number would be very unlikely in an inherited quality; it might be found in some situation where children tended to react against their parents' views.

The graph then in Figure 22 does not show how the children's intelligence grew; it shows how far *the order in which the children placed* tended to resemble the order in which their actual mothers placed. It will be seen from the figure that when the children were 2 years old, the correlation was 0; that is, at this stage it would be impossible to tell which children had the more intelligent mothers. However the correlation grows steadily as the years pass. By the time the children are 14, the correlation has reached 0·4, very nearly the figure for normal homes. The correlation with the educational level of the adopting mothers is negligible.

One might be tempted to conclude that intelligence goes its own way, completely determined by heredity and uninfluenced by how a child is treated. This however would completely ignore a factor that does not appear in the graph. The average I.Q. score of the mothers was 86; the average score of the children was 106. So our conclusion that the above-average mother was likely to produce an above-average child still stands, but the two averages in question are separated by 20 points. The effect of placing the children in good homes had not been to wipe out their inherited differences, but it did result in lifting the performance of the group as a whole. (Owing to a technical statistical effect, known as regression to the mean, the homes can only claim to have produced 13 points increase rather than 20; that however is still a considerable achievement.)

Figure 23 is taken from a paper by Moss and Baer in the journal *Growth* for 1956. The graph shows how the weight of a rat's brain compares with the weight of its body at various stages of growth. A sharp bend appears in the graph, and shows that a sudden change has occurred in the mechanism of growth. Before the bend, the graph is steep; the brain is racing to reach its full weight as quickly as possible. After the bend, the graph is almost flat; the body is continuing to grow, but the brain has reached the size required, and shows hardly any increase. You may notice that the

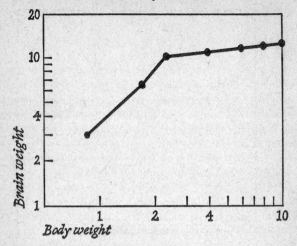

Figure 23

numbers on the scales in this figure are not evenly spaced. A
special type of graph paper is being used, which will be described
in our chapter 'Making the Curved Straight'.

A sharp bend also appears in Figure 24. This graph is based on
a paper by Deevey in the *Quarterly Review of Biology* for 1947
with the title 'Life Tables for Natural Populations of Animals'.
Deevey in turn drew on figures provided by Murie in his paper
The Wolves of Mount McKinley. The graph shows the chances of
survival for mountain sheep in a region where there are wolves.
The graph indicates that, of 1,000 sheep, about 800 would reach
the age of 1 year; further points on the graph show how many
would reach the ages of 2, 3, 4 . . . years, until finally only 6 sheep
reach the age of 12 years and only 3 the age of 13.

It will be seen that the graph begins with a very sharp descent,
indicating that the first year of a sheep's life is a very dangerous
period. Then the graph goes almost flat; once a sheep is past its
first birthday, it has very good prospects of standing with the
herd and defying the wolves or running to higher ground where
the wolves cannot follow. Then again gradually the graph becomes steeper; as the sheep becomes aged, its chances of being

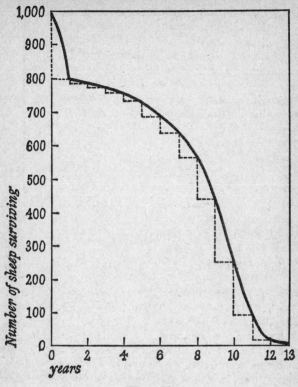

Figure 24

caught once again become worse. There is an apparent paradox at the end of the graph: after the 11th year the graph seems to flatten out once more and particularly between 12 and 13 the graph seems about as flat as it was between the ages of 1 and 2. Is extreme old age a kind of insurance policy for a sheep? The reason for this flatness however is not that the sheep are protected; it is that there are so few of them. Between 1 and 2 only 12 sheep out of 801 die; by the age of 12, there are only 6 sheep left altogether and the wolves get 3 of these in the following year. With only 6 sheep it is impossible to get a drastic plunge like that between ages 9 and 10, when 187 out of 439 fall to the wolves.

The Search for Pattern

It may be of interest to give the actual figures, so that you can see how these reflect and confirm the conclusions drawn from the graph. The symbols used are as follows; x denotes the age of the sheep, l_x denotes the number of sheep, out of 1,000, still alive at that age, and d_x the number that will die in the following year. Thus l_7, for example, denotes the number of sheep that survive to the birthday that makes them 7 years old; d_7 indicates the number of these that will die before reaching the age of 8.

x	l_x	d_x
0	1000	199
1	801	12
2	789	13
3	776	12
4	764	30
5	734	46
6	688	48
7	640	69
8	571	132
9	439	187
10	252	156
11	96	90
12	6	3
13	3	3

In Figure 24 the dotted lines form a kind of staircase on which the graph rests. This staircase helps to bring out the variation in the steepness of the curve. The level lines are all of the same length; each represents the passing of one year. But the upright lines are of different lengths. They show the decrease in the number of sheep surviving; in fact, their lengths are given by the numbers d_x. These numbers give us a way of measuring how steeply the curve is falling.

The staircase here is very different from the staircase we met in the previous chapter in connexion with the linear function, where every step had the same height (as in the stairs usually found in houses). It is important not to carry over to curves ways of thinking that only apply to straight lines, as beginners in algebra are sometimes apt to do. When dealing with any formula, one should consider the sort of graph it has, and at the very least be clear whether the graph is straight or not. For instance, the formula $y = 6x$ does have a straight graph. When we examine its table

x	0	1	2	3	4
y	0	6	12	18	24

we see the same increase, 6, as we go from each number in the y row to the next (and the graph in fact behaves properly in between these values). The particular equation $y = 6x$ was chosen because of its relation to a well-known puzzle, which runs as follows. A steel band has been put round the equator of the earth. It is then found necessary to raise this band a yard above the surface of the earth, all the way round the equator. The equator is about 25,000 miles long. Naturally, the steel band will have to be made longer, since all the way round this vast circle it is being pushed a distance of one yard further away from the centre of the earth. The question is – how much longer would the band have to be made?

Apparently the impression this puzzle is trying to create in the minds of listeners is that, since the equator is so long, an enormous amount of extra steel will be needed. However, this is not so, nor is it necessary to do any arithmetic involving the exact radius of the earth. The circumference of a circle is, roughly, 6 times as long as the radius of the circle. (Actually, the number is nearer to 6·28, but for the principle involved it is quite sufficient to use the round number.) Accordingly, every time the radius of a circle is increased by a yard, the circumference is increased by about 6 yards. It does not matter whether the circle is a large one or a small one. All the steps are of the same size. If the circle happens to be the equator, an extra yard on the radius will still mean an extra 6 yards or so on the circumference. Accordingly, about 6 yards more steel band would be required.

The equal steps, which we observe for the formula $y = 6x$ and for other linear equations, do not occur in general. As an example it is sufficient to consider the formula $y = x^2$. If we take in turn 0, 1, 2, 3, 4 for x the numbers we obtain for y are 0, 1, 4, 9, 16 and the steps from each number to the next here are certainly not equal. One could illustrate the difference in the following way. Suppose some signal, such as a drum beat or the click of a metronome, occurs at regular intervals. A large supply of tiles is to hand, and in order to illustrate the meaning of a formula you are given the job of seeing that y tiles are on the table when the xth beat of the drum sounds. If the formula to be illustrated is $y = 6x$, you work at a steady rate. Between each beat and the next, you have to put 6 more tiles on the table. If the formula is $y = x^2$, you have to have 1 tile on the table in time for the first beat, 4 in time for the second, 9 in time for the third, and so on. These are the square numbers, so we might imagine that you actually arrange the tiles in squares. In each interval, you have to put on an extra border, so as to enlarge the side of the square by the width of one tile. But as the squares grow, the border becomes longer and longer, and you have to work harder and harder to get the tiles in place in time. It is perhaps not surprising that in mechanics the formula $s = t^2$ turns up in connexion with accelerated motion.

GRAPHS WITHOUT EQUATIONS

As we have mentioned, whenever you are dealing with a formula or an equation, it is good to have some idea of the appearance of the corresponding graph. This will give you an indication of the kind of situation you are dealing with and may warn you if an error has crept into your calculations. But it is also possible to work with graphs without considering their equations at all. The graph we had earlier, relating success at school to the degree of anxiety felt, was such a graph. This graph was drawn by a free sweep of the hand; it did not represent an equation, or even exact experimental data; it was simply meant to illustrate an idea, a suggestion, 'Probably the effect of increasing anxiety is something like this.'

Figure 25 shows a graph that arose in connexion with a very crude home-made radio set. The tuning knob had a pointer on it, but there was no scale to indicate the angle at which this pointer stood; the angles had to be estimated by eye. Three stations were heard and identified by announcements; one on 590 kilocycles seemed to come in at a pointer setting of about 85°, another on 740 kilocycles around 130°, and a third on 860 kilocycles about 175°. A fourth station came in at an estimated 105° but no announcement of its identity was heard for a long time. Accordingly this graph was drawn in an attempt to find out which station it

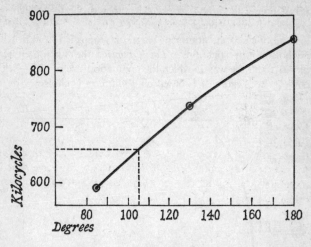

Figure 25

was. The pointer angles were plotted as x and the kilocycles as y, and a curve was drawn freehand through the points representing the three known stations. The dotted lines then indicated that a pointer angle of 105° should correspond to a frequency of 660 kilocycles. A list of nearby stations showed that there was a transmitter on 680 kilocycles, and so it was guessed that this was the fourth station, which in fact it turned out to be – perhaps surprisingly, in view of the very shaky data used. The procedure used here is one that can very frequently be applied. The problem is known as that of *interpolation*, that is, given several points on a graph (or the corresponding data) to find a point on the graph somewhere between them. The method of course represents a kind of guesswork; wildly different curves could be drawn through three points. One could for instance have drawn an S-shaped curve, rising to a great height between the first two points, and falling to a great depth between the second pair. Experience of radio circuits suggests that such a curve is not to be regarded as a serious possibility. Thus, mathematically, there is no meaning in talking about *the* curve through three given points; in an interpolation problem we are looking for a curve that goes through

51

(say) three points, and has certain other properties which we, consciously or unconsciously, require of it.

A more elaborate argument based on graphs is found in A. J. Nicholson's investigations of *Lucilia cuprina*, the Australian sheep blowfly, and the way in which the population size of these flies is determined. Figure 26 is based on a graph in Nicholson's paper

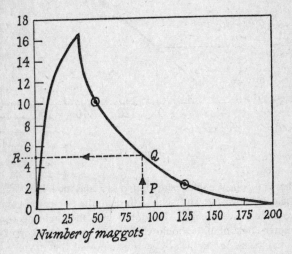

Figure 26

to the Eighth International Congress of Entomology at Stockholm in 1950. Nicholson had a number of test-tubes in each of which 1 gramme of meat was placed as food for newly hatched blowfly maggots. Into some tubes only a few maggots were introduced, into others many. It was then observed how many adult flies emerged in each tube. The graph shows the results. So long as the number of maggots in a tube did not exceed roughly 30, the more maggots there were, the more flies resulted. But beyond this point, which corresponds to the peak of the graph, increasing the number of maggots decreased the number of adult flies. The maggots competed for the food and many of them could not obtain sufficient to survive. If 200 or more maggots were placed in a tube, they all starved and no adult flies at all were found.

What will happen now if succeeding generations grow up in these conditions, that is, if for each generation 1 gramme of meat is provided, but apart from this the numbers are left to look after themselves? The adult flies produced in one generation will lay eggs and thus determine the number of maggots in the next generation. For the sake of illustration we will suppose that each fly lays 25 eggs. (If we wished to take account of the sex life of the blowfly, we would have to re-word this; suppose that on the average 25 eggs are laid for every adult fly that emerges.) If it happened that there were 100 maggots at the start, the graph shows that 4 adult flies should survive. Each of these laying 25 eggs would give 100 maggots in the next generation and so theoretically the population should continue for ever at this figure. The situation however is unstable. Suppose there happened to be only 90 maggots at the start – slightly less than the number needed for a steady state. From the graph, there will be 5 adults resulting.* These 5 adults will lay 125 eggs. From 125 maggots the graph shows that only 2 adults emerge. These in turn will lead to 50 maggots, of which 10 will survive as adults. This is the largest number of adult flies so far and it is fatal to the fly population, for 10 adults will produce 250 maggots and a famine ensues which none will survive. It can be shown that, in the circumstances described, starting from any number except 100 maggots leads to this result sooner or later – total annihilation. We will consider later why it is that this process has not made blowflies extinct. First however it is worth observing that we can avoid the arithmetic used above and represent the whole process graphically. Figure 26 showed the number of flies to which a given number of maggots led. But we can also show the next stage of the process, the maggots produced when these adult flies lay eggs, by a graph. As each adult produces 25 maggots, the graph is the straight line shown in Figure 27. Let us call these two graphs the survival graph and the egg-laying graph respectively, and observe how it is that 90 maggots in the first generation lead to 125 in the second. We begin with the survival graph. The point P corresponds to 90 maggots; we go up from it to Q and then across to R,

* Nicholson's paper gives the graph, but not the exact numbers used in drawing it. I have taken the liberty of interpreting his graph in such a way that it produces an arithmetically simple illustration.

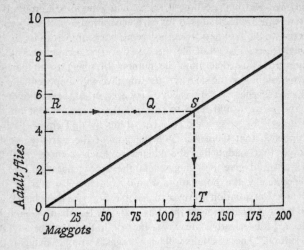

Figure 27

to find that 5 adults result. Then, on the egg-laying graph, we start at R, go across to the point S and descend from it to T, the point that represents 125 new-born maggots. Now in going from R to S we pass through the point Q that we used in the earlier part of the construction. Accordingly, going from Q to R and back again is in certain respects a wasted journey. We can find the point S more quickly and economically by going directly from Q to S. In the same way, part of our downward journey from S to T will prove to be wasted, for on arriving at T the next thing we do will be to go up from T until we meet the survival curve, and discover that of the 125 maggots only 2 survive to become parents. It would therefore save effort if we descended from S only until we hit the survival curve. If we were to follow this procedure, only one fiddly operation would remain. We would have to mark on the egg-laying sheet the point Q that we first found on the survival curve, and we would have to mark on the survival sheet the point S which was originally determined on the egg-laying sheet. A natural way to avoid this would be to have both graphs drawn on the same sheet of paper, as in Figure 28. Starting from P and following the arrows we can follow the whole process. The point P represents 90 maggots; we rise from P to Q on the survival curve, and see that 5 become adult; going across to S on the egg-laying line, we find S is above the mark 125, so 125 maggots come from the 5 adults; U on the survival curve indicates that only 2 of these become adult – and so we go on, to V and W, and finally the line WZ hits the egg-laying line right off the diagram, at a point where no maggots survive.

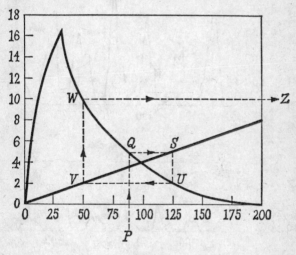

Figure 28

This procedure has been explained in some detail, which I hope readers will not find unduly wearisome, as there is a good deal to understand in Figure 28 and it is not wise to rush things. Incidentally, the procedure used here is also found in applications of mathematics far removed from biology, for example in the theory of solving equations with the aid of an electronic computer.

If Nicholson's experiment described above gave the whole truth about blowflies, there would be no blowflies. The blowfly population would oscillate, going above and below its steady value, and eventually destroying itself by an over-population famine. Nicholson points out that he puts the maggots into the tube at one time, whereas in nature the different generations would overlap. Again, in a later paper (*Australian Journal of Zoology*, 1954) he points out that the adult blowfly can live on water and sugar, but can only produce eggs if it gets a sufficient amount of protein. It is therefore possible to have a very high population of adults and yet for no eggs to be laid. Figure 29

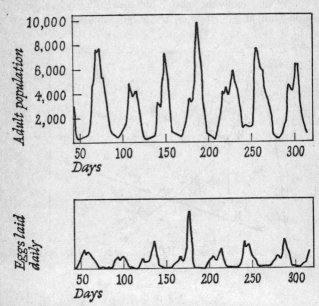

Figure 29

shows the oscillations Nicholson was able to observe in a large colony of blowflies. It will be seen that periodically there is intense overpopulation and almost no eggs are laid; the population then dwindles to a very small number; there are now only a few, but well-fed adults; they start to lay eggs and the whole cycle repeats itself.

Some general considerations perhaps emerge. On the one hand it is clear that the mechanisms of nature are complex. Nicholson suggests in his 1950 paper that there may be circumstances in which a campaign of killing insects may produce an actual increase in their numbers. Again, in television commercials one can observe a somewhat simplified view of medical science. One is urged to plaster oneself with some product because it 'destroys germs instantly'. That it might thereby create ideal conditions for a population explosion of new and perhaps more troublesome germs is not mentioned.

Nature is complicated and it is not wise to interfere with it rashly. On the other hand, there is equally little basis for the view that Nature is a kindly, all-knowing mother, and that we ought not to interfere with nature at all. Recurring phases of famine may keep the blowfly population between certain limits, but this is not a mechanism we would welcome for preventing human over-population. If human life is to be endurable we have to interfere with the universe, but realizing as we do so that it is a complicated and dangerous undertaking.

CHAPTER THREE

Making the Curved Straight

BEFORE we can deal successfully with any situation, we have to understand the laws that govern its development. In simple situations, the laws may correspond to formulas in algebra. In physics, which is the most regular of all sciences and the closest to mathematics, this has been recognized for several centuries. However even in other departments of inquiry, where the processes are much more involved and complicated than those of physics, we can sometimes identify features capable of simple algebraic description.

Graphs are a particularly convenient way of demonstrating – or discovering – a simple law underlying a set of experimental data. As we saw in Chapters 11 and 12 of *Vision*, a linear equation such as $y = 3x + 2$ or $x = 5$ corresponds to a straight-line graph, and in fact every straight line has an equation either of the form $y = mx + c$ or of the form $x = k$. All other equations give curves.

Now it is easy to recognize a straight line, but it is hard to distinguish one curve from another. If you were to plot the graphs of $y = x^2$, $y = x^3$ and $y = x^4$ say for values of x between 5 and 10, three curves would result, having the same general characteristics. Relatively few people, on being shown one of these curves – without the others for comparison – would be able to say which it was. Of course, there are various procedures involving calculation by which one can fit an equation to a given curve. However calculation is tedious as compared with a process that simply involves plotting a graph and looking at it. Further, data often contain small errors. For instance, when a length is measured, an experimental error is almost certain to be produced. Such errors produce small changes in the positions of the plotted points. One can often see that a collection of points lie nearly in a straight line, and this can be checked with the help of a ruler or a stretched thread. It is considerably harder to reach the same conclusion purely by calculation.

Now of course the equation $y = 3x + 2$ gives a straight line only if it is plotted on properly made graph paper, with evenly spaced divisions. If the spacing were uneven, $y = 3x + 2$ would cease to give a straight line. A straight line would then indicate some other kind of formula or rule. By playing around with the spacing, we can produce graph paper specially suited to certain types of law.

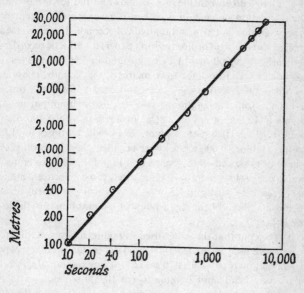

Figure 30

Figure 30 shows the world records for running various distances. The distances are in metres, the times in seconds. Records made after 1 October 1965 are not shown in the source I used. It is very noticeable that the points plotted lie very nearly in line on this graph paper.

Now if there were a simple formula, of the type $y = mx + c$, behind these data, the numbers would go up by equal steps. To run no distance takes no time; to run 5,000 metres takes just over 800 seconds. If the graph (on ordinary squared paper) were a straight line, each extra 5,000 metres would mean approximately

800 seconds more. But this is not so. Naturally, runners go more slowly over the longer distances. The record for 30,000 metres is about 1,000 seconds more than that for 25,000 metres. If you care to plot the records at intervals of 5,000 metres, you will obtain a graph that is slightly curved. It is in fact remarkable how slight the curvature is – far less than one would guess. However, as the difference between the 800 seconds and the 1,000 seconds in the calculation above indicates, there is a curvature apparent even when we think in terms of hundreds of seconds. If we assumed a straight-line law (on ordinary graph paper) for athletic records, on the basis of 30,000 metres being covered in 5,641·8 seconds, we would predict a record for 1,000 metres of 188 seconds. The actual record is 136·2 seconds, so we are more than 50 seconds out.

I made some calculations, based on the assumption that the points in Figure 30 lie on a straight line joining the 100 metres point to the 30,000 metres point. The predicted times for 1,000 metres and for 10,000 metres were both less than 10 seconds out. I have not checked all the points on this graph, but the change of graph paper does seem to have yielded some improvement. An exact result is hardly to be anticipated; at any moment an old record may be broken, and a point of the graph move a little to the left.

It is evident that the scales used in Figure 30 are very different from those used in the usual squared paper. On ordinary paper, if you had marked 10 and 20 seconds at any reasonable distance apart, you would not be able to get 10,000 seconds on to the paper at all. This incidentally is one of the advantages of the scale used in Figure 30; quite apart from getting certain laws to appear as straight lines, it allows us to show a far greater range of numbers than we otherwise could. These scales are sometimes used purely for this reason. But they also have a remarkable effect in turning curves into straight lines. Each of the graphs $y = x^2$, $y = x^3$, $y = x^4$ becomes a straight line when this type of graph paper is used.

The purpose of this chapter is to describe two kinds of graph paper that can be used to identify laws behind data. The first is used for things that grow – or shrink – by the compound interest law. The second is the paper we have already used in connexion with athletic records.

COMPOUND INTEREST OR ORGANIC GROWTH

In order to construct our first kind of graph paper, we begin by considering what happens when a population grows according to the compound interest law. If a population grows at a steady rate of 2 per cent a year, it will double in 35 years. If we wait another 35 years, it will double again. If it continues to grow in this way

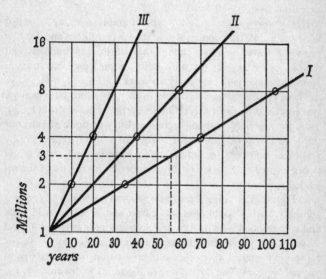

Figure 31

for 10 periods of 35 years (which is perhaps unlikely), it will double 10 times, and will end up 1,024 times as large as it began. Thus there is a simple relation between the number of years that pass and the number of times the population doubles.

The graph paper used in Figure 31 seems at first glance to be the usual squared paper – and in fact you can make this graph paper from the conventional squared sheet. The difference comes in the marking of the numbers. The numbers on the upright axis are not the usual 0, 1, 2, 3 . . . but 1, 2, 4, 8, 16 . . . That is, each upward step corresponds to a doubling. The line marked I corresponds to the situation discussed in the previous paragraph, a population that doubles every 35 years. It begins at 1 million. The plotted points show 2 million after 35 years, 4 million after 70 years, 8 million after 105 years – and these points lie on a straight line.

At whatever rate of compound interest a population may grow, there will be a certain length of time in which it doubles. Graph II shows a population that doubles every 20 years, graph III one that doubles every 10 years.

Exploratory exercise. On the type of graph paper we have just been using plot (i) a population that starts at 2 millions and doubles every 10 years (ii) a population that starts at 4 millions and doubles every 10 years (iii) a dwindling population, that starts at 16 millions and is halved every 10 years.

There is a certain arbitrariness in the procedure by which we constructed our graph paper. On the upright axis we wrote 1, 2, 4, 8, 16, the numbers doubling each time. But why double? Why not treble and write 1, 3, 9, 27, 81? Why not multiply repeatedly by any other number? In fact we could perfectly well make graph paper using 3 or 5 or 1·1 or any number larger than 1 as the repeated multiplier.

In Figure 31, graph I shows a population growing at 2 per cent annually. Such a population, as we have seen, doubles in 35 years and is multiplied by 4 in 70 years. Somewhere in between 35 and 70 years, it must be three times its initial size. Actually tables of compound interest show that our population trebles its size in about $55\frac{1}{2}$ years. The dotted line in Figure 31 is drawn from the point representing $55\frac{1}{2}$ years, and is carried across to meet the upright axis. The point where it meets the upright axis accordingly must represent the population after $55\frac{1}{2}$ years, that is, 3 millions. Accordingly the number 3 is marked at that point. In the same way, by using a table of compound interest, we could find at what times the population passed the 5 million, 6 million, 7 million marks, and use this information to mark 5, 6, and 7 on the upright axis. Thus we could mark every number, and not only the powers of 2, on that axis.

In practice, we do not have to go through all this work whenever we want to spot a compound interest law, because we can buy commercially prepared graph paper with the numbers already shown in the proper places. Figure 32 gives a simplified impression of such paper. Actually, the graph paper has many more lines than those shown in this figure.

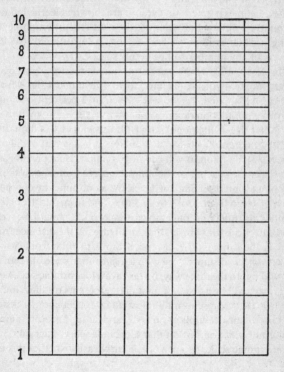

Figure 32

Various names are used for this kind of paper. If you were buying it, you might have to ask for arith-log paper, log-linear, or semi-log paper. Here *log* is short for *logarithm*. Logarithm is a long word for a simple idea. The spacing of numbers we used when we marked 1, 2, 4, 8, 16 . . . at equal intervals constitutes a *logarithmic scale*. In Figures 31 and 32 the upright scale is of this kind, but the horizontal scale is of a familiar kind, like the markings on a ruler.

Hence the name *semi-log*; there are two axes, upright and horizontal; we could have used the logarithmic scale, 1, 2, 4, 8, 16 . . ., on both axes, but we only went half-way to this. As for the name *arith-log,* this is because we have the usual arithmetical scale (even spacing) on one axis, and a logarithmic scale on the other.

It sometimes happens that graphs are shown in learned journals without the author bothering to say what type of graph paper is being used. It is however easy to spot a logarithmic scale. The powers of any number will be evenly spaced. If you see, 1, 2, 4, 8, 16 or 1, 10, 100, 1,000 marked at equal intervals, you can be pretty sure that you are dealing with a logarithmic scale. A further test can be applied. As we saw earlier, at 2 per cent compound interest a population doubles in 35 years. It does not matter what the initial size of the population is. If it began at 1 million, it will end at 2 million; if it began at 3 million, it will end at 6 million; if it began at 5 million, it will end at 10 million. Now on a logarithmic scale, the spacing of the numbers is read off from a table of compound interest. Since it takes the same time, at compound interest, to go from 1 to 2, from 3 to 6, and from 5 to 10, the corresponding intervals on a logarithmic scale should be equal. You can check for yourself that, in Figure 32, the distance from 1 to 2 is the same as the distance from 3 to 6 or from 5 to 10. Wherever this distance occurs on a logarithmic scale, it represents a doubling. In the same way, if you have a logarithmic scale before you, there will be a distance on it that represents trebling, another distance that represents multiplication by 10, yet another distance that represents multiplication by 7, and so on for every number. Accordingly, if you suspect that some unlabelled diagram is using a logarithmic scale, you can check fairly easily whether this is so.

SOME APPLICATIONS

We introduced semi-log graph paper by considering a population growing according to the compound interest law. Growth of this kind does actually occur. On a small scale it can be observed when bacteria are provided with plenty of food, the temperature and other living conditions being maintained in a state congenial to them. On a larger scale, we can observe such growth, over certain periods of time, for human populations. Figure 33 shows the

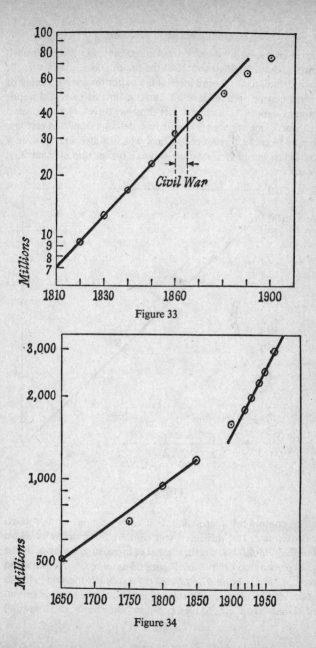

Figure 33

Figure 34

growth of the population of the United States in the nineteenth century. Between 1810 and 1860 the graph is essentially straight and indicates a compound interest law; after that, a slackening of growth can be seen. Figure 34 shows estimates of world population at various times in the last three centuries. Here two lines are drawn, one showing an approximate compound interest law from 1650 to 1850, the second line, which is steeper, showing a similar law in our own century, but at a higher rate of interest.

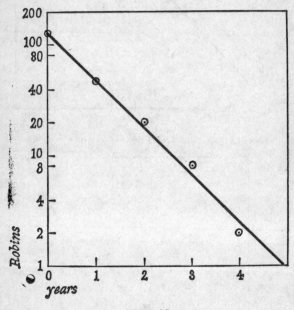

Figure 35

The compound interest law is usually referred to as the exponential law, and this name will be more appropriate when we consider populations which, instead of increasing, decrease by the same percentage each year. Figure 35 is based on information given in Lack's book *The Life of the Robin* (Witherby, 1943). It follows the fortunes of 129 robins, and shows the number surviving as the years pass. The way in which these points cluster around a straight line indicates that in each year a robin's chance of dying stays practically the same. This contrasts strongly with the mortality curves for human beings and for mountain sheep, both of which have a high mortality rate in infancy and in old age, with a very low rate in between. Many authors have maintained that a constant mortality rate is typical for birds, not only for robins. This has been questioned, on the grounds that it is very difficult to establish just how many wild birds are surviving at any time; on this view, the graph corresponds to the method of observation rather than to the actual experiences of birds.

Studies of bird mortality require long and carefully planned observation. However, the process of mortality at a constant rate can be simulated by a simple experiment, which anyone can do in a short time without even going out of doors. Imagine a game played by the following rules. The players throw dice. Any player who throws a 6 becomes dead – he is out of the game. The survivors throw again, and continue throwing until everyone is out of the game. A record is kept of the number who survive each round. One could think of each round as corresponding to a year; each player represents a creature, whose chance of dying in any year is 1 in 6. (The game birds play is considerably more dangerous; their chance of dying within a year is more like 4 in 6.) Of course, you need not have actual players. You can throw the dice yourself, and keep a record of how many sixes are thrown in each round. We are only concerned with how many players are eliminated, and not with which particular players these are. The quickest way to carry out the experiment is therefore to have half-a-dozen or more dice, make the appropriate number of throws for each round, observe how many sixes occur, and record how many players survive that round. Figure 36 shows the results I obtained when I followed this procedure, supposing the game to begin with 30 players. The straight line shows the way the population of players would decline if *exactly* one-sixth were eliminated at each round. Clearly, one cannot expect the experiment to reproduce this line exactly. It might happen that 5 players, one-sixth of 30, would succumb in the first round, and leave 25. It is however impossible that one-sixth of 25 should drop out in the second round, since this would involve a fraction of a player. Apart from this, there will be fluctuations due to the fall of the dice. Only 1 player remained after 16 throws, but he was fortunate; a 6 did not turn up until there had been 14 more throws of the dice. Anyone who has played ludo as a child or with children knows how much and how annoyingly the fall of dice can vary. When you want a 6 you may get it at the first throw or you may make a dozen or more throws without getting it. In fact, there is a 10% chance that, in 12 throws of a die, you will not get a single 6. The fewer players there are,

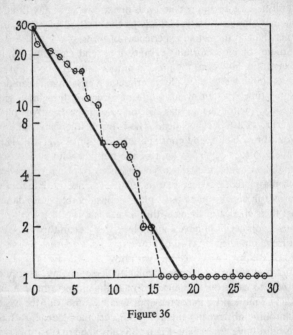

Figure 36

the more important fluctuations due to chance become. Figure 36 is instructive, because it relates to a situation in which we know exactly what is happening. If we assume the dice are true, then we have a situation in which the chance of dying in a year is exactly steady at 1 in 6. The figure shows the kind of deviations from a straight line that can be expected in this situation, and gives us an inkling of the irregularities likely to occur in an actual field study.

Figure 37 shows exponential decrease occurring in a medical

investigation. Most human beings have in their blood plasma a substance known as ACAH. About one person in 100,000 is completely lacking in this substance. People who lack ACAH seem to be perfectly healthy. However ACAH seems to be important for recovery from commonly used anaesthetics; serious consequences can occur if an anaesthetic is given to an individual

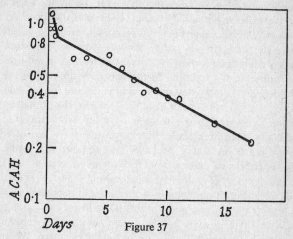

Figure 37

(© 1967 by the American Association for the Advancement of Science)

who has no ACAH. One can deal with this situation by giving a transfusion of normal plasma; the question then arises – how long will the ACAH from this normal plasma remain in the patient's bloodstream? Figure 37 shows how the ACAH decreases as time passes. There is a rapid decrease at first, probably due to the ACAH spreading through the body. There is then a more gradual decline. The straight line indicates that this is essentially an exponential decrease. Further details of this experiment will be found in a paper by Jenkins, Balinsky, and Patient in *Science*, Vol. 156, 30 June 1967, pages 1748–9.

C. A. B. Smith, on page 146 of his useful book *Biomathematics* (Griffin, 1954) mentions an interesting medical use of the exponential law. It was observed in the years 1916–18 that the area of a wound decreases exponentially as time passes and the wound heals. If a bend appears on the graph of area against time, on semi-log paper, this probably indicates that the wound has been re-infected.

POWER LAWS

We will now consider a second kind of graph paper that uses logarithmic scales. This is in fact the paper used at the beginning of this chapter to plot athletic records. This paper is known sometimes as *full logarithmic* paper (in contrast to semi-log paper) or as *log-log* paper (in contrast to arith-log paper). With semi-log paper we had the familiar, even-spaced scale horizontally, and a logarithmic scale vertically. In the paper we are now going to consider, both horizontal and vertical scales are logarithmic.

For the actual analysis of data, the commercially printed paper is almost essential, since there is so much work involved in making a logarithmic scale. But for an understanding of what log-log paper does, it is probably best to start with ordinary squared paper and make a simplified, home-made version of log-log graph paper. How to do this is shown in Figure 38. We draw a horizontal and a vertical axis. Where these lines cross is marked 1 on both axes. As we go from this point to the right along the horizontal axis, we mark 2, 4, 8 . . . , doubling at each step. As we go to the left we mark $\frac{1}{2}, \frac{1}{4}, \frac{1}{8} \ldots$, halving at each step. We mark the vertical axis similarly, doubling as we go up, halving as we come down.

To see the effect of this paper, let us plot $y = x^2$ on it. We first make a table, in which of course we include only the numbers marked on the paper, thus –

x	$\frac{1}{4}$	$\frac{1}{2}$	1	2	4
y	$\frac{1}{16}$	$\frac{1}{4}$	1	4	16

These points are plotted in Figure 38. It will be seen that – as promised near the end of the opening section of this chapter – the graph of $y = x^2$ is now a straight line.

The best way to get a feeling for the effect of log-log paper is to work a series of exercises such as those given below. On a moderate sized sheet of squared paper, there is plenty of room to mark the numbers from 1/256 to 256. The exercises should be worked on paper where this has been done.

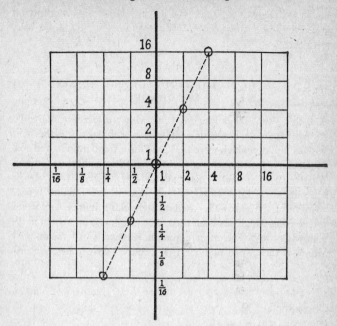

Figure 38

It is recommended that any reader, who is not already familiar with the use of log-log paper, should work these exercises before reading further. The exercises are not difficult, and the experience of doing them conveys an understanding of log-log graphing far better than any verbal explanation. It will be seen that the nature of each equation is clearly reflected in the appearance of the corresponding graph.

Exercises

1. Plot, on one sheet of paper, the following graphs: $y = x^2$; $y = 2x^2$; $y = 4x^2$; $y = \frac{1}{2}x^2$; $y = \frac{1}{4}x^2$.
2. Plot, on another sheet of paper, the following graphs: $y = x$; $y = 2x$; $y = 4x$; $y = \frac{1}{2}x$; $y = \frac{1}{4}x$.
3. On yet another sheet of paper plot: $y = x^3$; $y = 2x^3$; $y = 4x^3$; $y = \frac{1}{2}x^3$; $y = \frac{1}{4}x^3$.

4. Similarly plot: $y = 1/x$; $y = 2/x$; $y = 4/x$; $y = \frac{1}{2}/x$; $y = \frac{1}{4}/x$.
5. Similarly: $y = 1/x^2$; $y = 2/x^2$; $y = 4/x^2$; $y = \frac{1}{2}/x^2$; $y = \frac{1}{4}/x^2$.
6. Similarly: $y = 1/x^3$; $y = 2/x^3$; $y = 4/x^3$; $y = \frac{1}{2}/x^3$; $y = \frac{1}{4}/x^3$.
7. Plot similarly: $y = \sqrt{x}$; $y = 2\sqrt{x}$; $y = 4\sqrt{x}$; $y = \frac{1}{2}\sqrt{x}$; $y = \frac{1}{4}\sqrt{x}$.

THE MATHEMATICS OF THE RUSSIAN LANGUAGE

The thing that takes longest, when you are learning to read a foreign language, is getting a sufficient vocabulary. It is therefore wise to find out which words are used most frequently and to concentrate on learning these. E. Steinfeldt has published a book, *Russian Word Count*. This analyses 400,000 words in various Russian works, and shows, among other things, how often each word occurs. Figure 39 is based on Steinfeldt's figures. As you can see, log-log paper has been used. The graph is constructed in the following way. The most common Russian word means *and*, which is thus word number 1. It occurs 14,576 times. Accord-

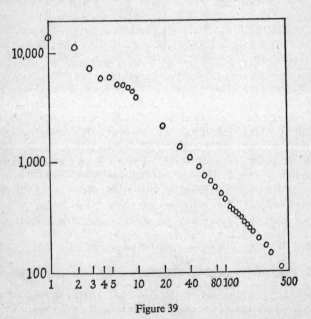

Figure 39

ingly we plot on the graph $x = 1$, $y = 14,576$. The fifth most common word means *he*, occurring 6,067 times; it is plotted as $x = 5$, $y = 6,067$. The other points are plotted similarly. It will be seen that after a little bit of initial wobbling, the points settle down into a good approximation to a straight line.

It is a rather striking fact that the commonest 22 words account for more than 25 per cent of the language. This suggests that, with very little labour, we ought to be able to get a good impression of what a passage in Russian is about. However, this is not so; the commonest words convey very little meaning, as may be seen by the following sample from a child's book in Russian, in which the 22 commonest words are translated, and all other words are replaced by blanks. '— — — — — into — . In — she — and — — — — — not — , but — in — to — . — was — : she — in — , — in — — — , and — . In this — — — — . — — was — , — him — — . She was — , and — her — — . — was — — , and — him — . — were not — , they — — — — .' Naturally, the language in a child's story is simple, and the 22 most frequent words account for 31 of these 86 words. But we still have not the faintest idea of what the story is. Actually it is the Three Bears – 'A girl went from home *into* forest. *In* forest *she* lost *and* began seek way home, however *not* succeeded *but* came *in* forest *to* hut' – and so on. The crude translation is an attempt to keep the number of words the same; English often uses two words where Russian uses one.

The knowledge of the commonest words of a language in fact does little more than enable you to tell which language is being spoken. It is a consequence of the distribution shown in Figure 39 that, while the first few words give you a surprisingly large proportion of the words in a book, the rate of advance becomes increasingly slow as you go on. Only 22 words give you 25 per cent; you have to go to 1,300 to get 74 per cent, to 2,000 to get 79 per cent, and to 5,500 to get 90 per cent. These figures refer to general literature. If your interest is limited to a single subject, you can of course manage with a smaller vocabulary.

THE SIZES OF CITIES

In most countries, the number of extremely large towns is small but there are many towns of moderate size. It is perhaps surprising that the law of distribution we observed for Russian words – and a similar law is found in very many languages – applies also to the distribution of city sizes. Figure 40 shows the distribution of population in Canadian towns, as reported in the census of 1961. There is of course a certain arbitrariness in the table; towns near to an expanding metropolis may retain legal identity long after

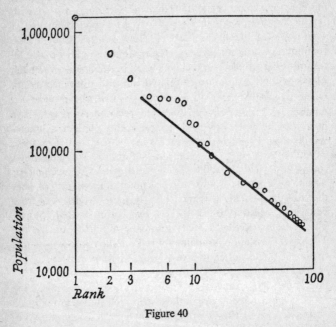

Figure 40

their boundaries have ceased to be visible to the naked eye. The figures are those recorded in the census, and reflect the legal rather than the physical situation. However this observation applies mainly to the larger centres, and in any case the graph shows irregularities for these. The approximation to the straight line is marked for the population distribution of the smaller towns. In this graph, the numbers on the vertical axis represent the population of a city. The horizontal number shows where the city comes in order of size. Thus Ottawa, the sixth largest town with a population of 268,206 has $x = 6$, $y = 268,206$.

In many other countries the distribution of city populations gives a straight line on log-log graph paper. See for instance John Clarke, *Population Geography* (Pergamon, 1965). A mass of information on this and other topics related to log-log graphing has been collected in G. K. Zipf's book, *Human Behavior and the Principle of Least Effort* (Addison-Wesley, U.S.A., 1949).

INCOMES

In 1896, Pareto published an observation on the distribution of incomes, usually referred to – both by those who believe in it and those who do not – as Pareto's Law. Pareto's Law attempts to connect two quantities – a certain amount of income, and the number of people who receive that amount *or more*. These quantities are plotted on log-log paper and, according to Pareto, should always give a straight line, and the steepness of that line should always be the same. Figure 41 is based on *Whitaker's Almanack* for 1963, and shows the distribution of incomes in U.K. in 1961. In plotting such a graph, care must be taken to remember the words *or more*. Thus Whitaker shows 5,000 with incomes over £20,000 and 20,000 with incomes between £10,000 and £20,000.

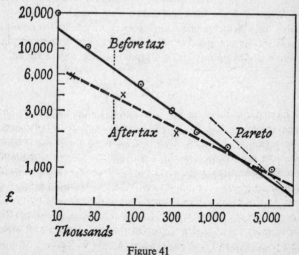

Figure 41

We have to add the 5,000 to the 20,000 to get the 25,000 who have £10,000 or more. Each step along the income table involves a further addition.

Pareto's Law claims validity only for the higher incomes before tax. Figure 41 agrees well enough with Pareto's contention that the graph should be a straight line. It does not agree with the view that in all societies the slope is the same. The slope predicted by Pareto would indicate a greater inequality of incomes than actually obtained. The income after tax, which after all is the most practical consideration, of course shows an even flatter graph.

Pareto's Law in its strongest form – the contention that in all societies the distribution of incomes is essentially the same – carries a strong tinge of fatalism. It suggests that, whatever we may do or attempt to do, certain factors inherent in human nature will keep things as they are and always have been. For industrialized nations at a certain stage of evolution Pareto's formula may give a realistic description. On the other hand in the traditional society of the Eskimos, the American Indians, the Pygmies, or the Australian aborigines it is doubtful if Pareto's Law could even be stated, let alone verified. Pareto's Law may apply to societies which, in the words of a young Asian visitor to the United States, have a lot of know-how and very little know-what-for. Whether know-how and know-what-for are mutually exclusive, or whether some future society may learn to combine them, remains a matter for individual judgement or faith or hope. Pareto's graphs do not settle this question one way or the other.

GROWTH

As any living creature grows, a certain harmony has to be maintained between the parts of its body. For example, if the distance from the elbow to the finger tips were three times the distance from the elbow to the shoulder, it would become impossible for the hand to touch the head. Insects could not be brushed off the face, or painful objects removed from the eye, or food brought to the mouth by using the hand. Yet all parts of the body do not grow at the same rate or in the same way. In man the brain has a high priority. At birth it accounts for one-eighth of the total weight; at maturity, less than one-fortieth. (See D'Arcy Thompson, *On Growth and Form*, Cambridge, 1942, page 185.) The brain practically reaches its final weight by the age of six years; the heart and the body generally continue to grow until the early twenties. Many separate processes are thus involved in growth; many laws operate; there are stages where one kind of development is replaced by another.

While the process of growth is evidently of immense complexity, much of the literature about growth is both attractive and simple. D'Arcy Thompson's book, which was originally published in 1917 and did much to stimulate the application of simple mathematics to biology, is eminently readable. So too is Julian Huxley's *Problems of Relative Growth* (Methuen, 1932). Few of the papers in *Growth*, a scientific periodical founded in 1937, are difficult to read.

Huxley's book made a contribution to the theory that is particularly appropriate to our topic of log-log graphing. The simplest animals, which consist of one cell only, reproduce themselves by splitting into two cells. The growth of higher animals depends on the same process, by which one cell becomes two. Thus growth occurs by the doubling of cells, and the rate of growth of any part of the body depends on the length of time between one doubling and the next. Thus the natural way to measure growth would be to see how many doublings occur in some interval of time. Now the number of doublings is precisely the measure we adopted in constructing our home-made graph paper. This suggests that we might use semi-log paper for graphing growth, with time evenly spaced along the horizontal axis, and a logarithmic scale vertically to measure the number of doublings. However, as Figure 42 shows, when we plot growth against time with semi-log paper, we get a very irregular curve. This graph shows how the whole body increases, between birth and 21 years of age, to about 20 times its initial weight, while the heart increases to $12\frac{1}{2}$ times its initial weight. Some of the jumps on the heart's graph, I would guess, may be errors of measurement. The heart is not an easy thing to weigh. But the variations in the rate of growth of the whole body are genuine – the very rapid growth in the first year of life, the spurt at adolescence, must appear in any adequate account of human growth.

The body is working hard at growing in the first year or two of

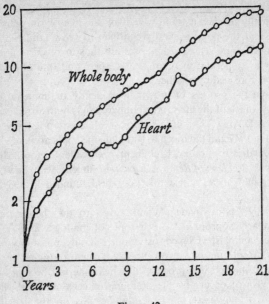

Figure 42

life, then it slackens off, but makes a spurt in adolescence. Huxley pointed out that we could overcome this irregularity by not comparing the body with time, or the heart with time, but comparing the growth of the body with that of the heart. Figure 43 shows the result of doing this. It is immediately apparent that the law relating the development of the heart to that of the whole body is much simpler than that relating either of them to time.

Figure 43 was drawn on log-log paper, but actually the heart's development closely parallels that of the rest of the body, and you would still get something very much like a straight line if you plotted the points on ordinary squared paper. To bring out the advantage of log-log paper, it is necessary to consider an organ that develops at a rate much different from that of the whole body. Huxley gives the example of the male fiddler-crab. When young, this crab has a claw whose weight is about 9 per cent of that of the rest of its body. The claw grows faster than the rest of

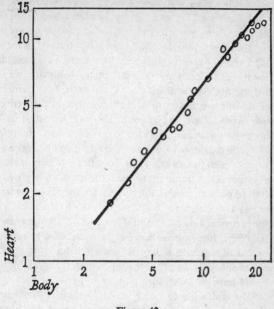

Figure 43

the body, and ends up being about 60 per cent of the weight of the rest. The early stages of the process are shown here:

Weight of rest of body	Weight of claw
mg.	mg.
57·6	5·3
80·3	9·0
109·2	13·7
156·1	25·1
199·7	38·3
238·3	52·5

The full table will be found on page 12 of Huxley's book. If you plot the points corresponding to these numbers on ordinary squared paper, a definite curvature can be seen. If however they are plotted on log-log paper, they cluster around a straight line. If the claw is plotted vertically, and the rest of the body horizontally, the line rises with a slope of 8:5 approximately; that is to say, the line rises 8 inches for every 5 inches across. This would indicate that (if the growth continued long enough according to this law) the claw would double 8 times while the rest of the body doubled 5 times. Accordingly we could prepare a cartoon film of the growth of the crab and its claw in the following way. We would draw the claw and the rest of the crab, both growing at compound interest, but at different rates, so that the periods of doubling would have the ratio 8:5 mentioned above. When we say 'growing at compound interest' this does not refer to time – for our data above do not tell us anything about the age at which crabs reach the various sizes – but *to the frames of the film*. As we go from each frame to the next, the main body grows by a certain fixed percentage and, the claw by another fixed, but larger, percentage. If we wanted to show the growth of the crab as it actually occurs in time, we should have to obtain data showing whether crabs, like human beings, grew particularly rapidly in infancy and adolescence. If they did, we would have to instruct the operator to speed up the projector appropriately at suitable times. We should have to make sure that we did not use this law for parts of the crab's life to which it did not apply. According to Huxley, this law applies for a considerable stage of development; after that another, very similar law applies.

TECHNICAL NOTES

So far, the work of this chapter has involved almost no calculation. In the main, the procedure has been simply to plot data on semi-logarithmic or log-log graph paper, and observe whether the resulting points lay, more or less, on a straight line. Some readers may be content with this, and indeed as it stands this approach is both of interest and of value. However, in scientific papers, the results of graphical work are often stated in algebraic form. It will be useful to indicate how this is done. Again, data are sometimes available that are far more accurate than what a graph can show. By measuring the steepness of a line drawn on graph paper we might, perhaps, get within a few per cent of the correct value. If our data happened to be accurate to one part in a million, this would be wasteful, and we would wish to replace the measurement by a calculation. It therefore seems useful to sketch briefly the properties of logarithmic graphing, and to mention something about the notation in which its results are sometimes presented.

The central idea is that of a logarithmic scale. As we saw with our home-made logarithmic graph paper, this scale is not difficult to construct. In the scales we made, doubling was the basic operation, and we marked 1, 2, 4, 8, 16 . . . at equal intervals. This is simple and convenient for explaining the *idea* of a logarithmic scale, but it is not very helpful for practical purposes, since there are numbers which lie between 4 and 8, between 8 and 16, and so on. We have no way of knowing where to mark these. This difficulty can be overcome if, instead of repeatedly multiplying by 2, we use a number closer to 1, such as $1 \cdot 001$ or $1 \cdot 000001$. Suppose we use the latter. Then we shall mark at equal intervals 1, $1 \cdot 000001$, $(1 \cdot 000001)^2$, $(1 \cdot 000001)^3$, and so on. This would be very tedious to do; we would only mark the number 2 after covering rather more than 693,000 intervals. By the millionth interval we would still not have reached 3, and after covering 2,300,000 intervals we should still be a little short of marking the number 10. In practice, of course, short cuts are found to avoid this labour. But in principle this procedure defines how the numbers should be spread out to make a logarithmic scale, and in fact this was the idea that led to the discovery of logarithms at the beginning of the seventeenth century – compound interest at a very low rate.

We have said that the numbers should be marked at equal intervals, but we have not said what the length of these intervals should be. And in fact, it does not matter. The interval could be an inch or a centimetre or a millimetre or a micron. Mathematically, it does not affect the validity of the scale. In practice, of course, it would be inconvenient to have graph paper that covered an acre or had to be read under a microscope.

Tables of logarithms are a familiar sight. One way of regarding a table of logarithms is as instructions for making a logarithmic scale. *The logarithm of a number specifies the distance at which that number should be marked on a logarithmic scale.* The logarithm of a number, *n*, is usually abbreviated to log *n*. A table could contain such information as log 1 = 0, log 2 = 0·301, log 3 = 0·477, log 10 = 1. This means that, if you are making a logarithmic scale you should mark 1 at distance 0, 2 at distance 0·301, 3 at 0·477, and 10 at distance 1, as shown in Figure 44.

There could really be lots of different tables of logarithms since, as we have noted, the size of the interval can be chosen at will. In fact, only two kinds of table will be met in practice. One is the table of Common Logarithms, in which the size of the interval is chosen in such a way that the number 10 is marked at the distance 1. The numbers given in the previous paragraph were taken from

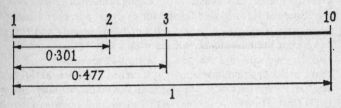

Figure 44

a table of Common Logarithms. The other table is that of Natural Logarithms, also called Napierian Logarithms, after the inventor Napier. The essential idea here to mark 1 at 0 and 1·000001 at distance 0·000001. The first table of logarithms ever made was one of Natural Logarithms. Natural Logarithms are still of considerable theoretical significance. My account of them here is slightly simplified. Further details, and an account of the use of the logarithmic scale in the slide rule, will be found in *Mathematician's Delight*, in the index of which several entries will be found under 'logarithm'. There too will be found a more leisurely discussion of points which in this present chapter are explained briefly or are noted without explanation.

FINDING THE SLOPE ON LOG-LOG PAPER

Our first example in this chapter was concerned with world athletic records. According to the source I used, the record for 100 metres was 10·0 seconds and for 30,000 metres was 1 hour 34 minutes 1·8 seconds, or 5,641·8 seconds. By examining the graph of these data, on the log-log paper used to make Figure 30, you can see that the time for 30,000 metres is somewhere around.

5,600 seconds. You cannot get closer than that; you certainly cannot detect that the time is 5,641·8 seconds. In this particular question, I doubt if extreme accuracy will help us, but suppose for some reason we did need to know, as exactly as possible, the slope of the line joining these two points. Tables of logarithms tell us, with considerable precision, where the points should be marked. As we saw a couple of paragraphs ago, the logarithm of 10 is 1, and on a logarithmic scale 10 should accordingly be marked at distance 1. The logarithm of 100 is 2, and so 100 metres will be marked at a height of 2 units above the origin. (It is necessary, when dealing with log-log paper, always to bear in mind the distinction between the number written at a particular place, say 100, and the physical distance from the origin, say 2 inches, at which that number is written.) Similarly, we have log 5,641·8 = 3·75142 and log 30,000 = 4·47712. These numbers give us the actual distances across and up the paper for the placing of our second point. The situation is thus as shown in Figure 45. It can be

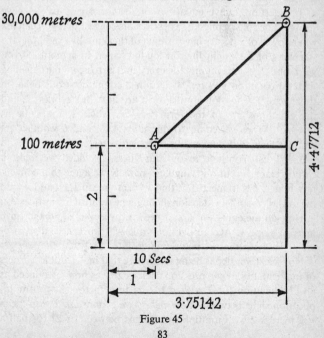

Figure 45

deduced from these that the distance AC is 2·75142 and the distance CB is 2·47712. The steepness of the line AB, by which we mean CB/AC, is thus 2·47712/2·75142, which works out to 0·9003. This steepness is thus very close to 0·9.

NOTATION

One could report the results of an experiment by saying that the data gave a straight line on log-log paper. This would indeed convey an important piece of information in a useful form. However another form is very frequently used, and this we will now consider.

On page 71 a number of exercises were given, in which various equations had to be plotted on home-made log-log paper. In Question 1, there were five equations, including $y = x^2$, $y = 2x^2$ and $y = 4x^2$. All the equations in this question were of the form $y = ax^2$ and on log-log paper all of them gave lines of steepness 2. In question 3, five equations of the form $y = ax^3$ all gave straight lines of steepness 3. These results suggest that you can tell what power of x occurs in a formula of this kind by examining the steepness of a straight line on log-log paper. In symbols, if you get a line that rises n units for every unit across, you are dealing with an equation $y = ax^n$. Question 2 strengthens our belief in this guess. The log-log graphs for it are lines of steepness 1, and the equations are of the form $y = ax$, which is the same as $y = ax^1$. When we come to questions 4, 5, and 6 we also get straight lines, but descending ones, with steepness -1, -2 and -3. We can continue to use our guess, provided we make a certain extension of our language, namely we agree to a convention that x^{-1} is to mean $1/x$, that x^{-2} is to mean $1/x^2$, and x^{-3} to mean $1/x^3$. Another extension of language is called for by question 7. Here all the equations are of the form $y = a\sqrt{x}$ and they give lines of slope $\frac{1}{2}$. Accordingly our rule will work here also, provided we agree that $x^{\frac{1}{2}}$ is to be interpreted as $\sqrt{x}$.

Now clearly a step is being taken here that may trouble learners of mathematics. An expression such as 2^4 was first introduced as an abbreviation for $2 \times 2 \times 2 \times 2$; four factors 2 with multiplication signs between. Now clearly we cannot use 2^{-3} to mean -3 factors 2 with multiplication signs between or $2^{\frac{1}{2}}$ for half a factor 2. Yet the use of negative and fractional indices has proved of great value in mathematics. Every writer of a mathematics textbook is aware of difficulty as he approaches the introduction of these ideas. The usual approach is to develop laws which hold for whole number indices, then to say that *if* these laws are to remain true when we consider negative and fractional values x^{-3} must mean $1/x^3$ and $x^{\frac{1}{2}}$ must mean $\sqrt{x}$. 'Yes,' says something in the pupil's mind, '*if* you are going to introduce these strange signs, and still use the same procedures, no doubt they have to be interpreted as you say. But maybe it is not justifiable to introduce them at all?'

The teacher of algebra would be in a stronger position if, after introducing these new signs, he went on to do something very clever with them and to solve problems with increased speed and power. Most algebra books do not provide any such demonstration of increased efficiency. They are not to blame; my impression is that, *within algebra*, negative and fractional indices do very little for you. The people who invented them, John Wallis and Isaac Newton in the mid seventeenth century, were not concerned with algebra. but were laying the foundations of calculus. Now in calculus this innovation really does work for you. In calculus, one is often seeking to find something called $\dfrac{dy}{dx}$, the meaning of which need not bother us at present. In simple cases, such as $y = x^2$ and $y = x^3$, one can find $\dfrac{dy}{dx}$ very easily, and the results are in accordance with the rule that if $y = x^n$, then $\dfrac{dy}{dx} = nx^{n-1}$. It is a great benefit to the learner of calculus when he realizes that, by using this same rule with $n = -3$ or $n = \frac{1}{2}$, he can deal with $y = \dfrac{1}{x^3}$ and $y = \sqrt{x}$, equations that look quite different from, and much harder than, the simple cases $y = x^2$ and $y = x^3$. So algebra teachers have to bear the brunt of introducing a rather novel idea that only becomes really profitable when calculus is reached.

Another interesting point is that most books first introduce the idea of fractional indices, which the learner does not want to accept anyway, and then build the theory of logarithms on top of

it. Now historically Napier's invention of logarithms preceded
Wallis' introduction of fractional and negative indices by nearly
half a century, so that the first table of logarithms was made by a
man who had never heard of fractional indices. Indeed, the seventeenth-century approach to logarithms was much simpler than the
one commonly used today. An idea worth exploring would be to
introduce logarithms in the seventeenth-century manner, and then
use the logarithmic scale to explain what fractional indices are.

The development would go like this. As we have already seen,
the construction of a logarithmic scale does not call for any
difficult idea. We imagine numbers marked at equal intervals
along a line, each number being 1·000001 times the number to its
left. The number 1 would be marked at the origin, but need not
represent the end of the scale, since we can write in the numbers to
the left of 1 by dividing by 1·000001 at each step. (In Figure 38
we used a procedure corresponding to this, but using 2 instead of
1·000001.) In this way each number finds its place on the line. I
have said 'each number', but of course this is not exactly true. Not
every number gets marked in this process, since each number
marked exceeds the number to its left by one millionth part. For
instance, if one of the numbers marked was approximately 7, the
next number to it would be approximately 7·000007. Now to four
places of decimals, both of these numbers are 7·0000, so the minute
gap between one number and the next would not cause us any inconvenience if we were making a table of four-figure logarithms.
(We will return to the question of logic involved here later.)

Now we have our numbers spread out along a logarithmic scale.
As we have already observed, for any number a, the numbers 1,
a, a^2, a^3, a^4, ... occur at equal intervals. Measuring from 1, we
find a^2 is twice as far away as a, that a^3 is three times as far away
as a, and a^4 is four times as far. We could then ask the learner,
'Mathematicians use the symbol, $a^{3\frac{1}{2}}$. Where do you think that
number will be found?' Surely, the answer would be, '$3\frac{1}{2}$ times
as far along the scale as a.' And in fact we could use as our definition of a^n that it is the number on the logarithmic scale n times
as far from 1 as a is. The beauty of this is that the learner has no
doubt that such a number exists. Numbers seem to be marked all
along the scale; he will surely find some number $3\frac{1}{2}$ times as far as
a. But he may very well have grave doubts about there being any

number which has 3½ factors *a* with multiplication signs between.

This definition lends itself very readily to exercises. Figure 46 here shows again the home-made logarithmic scale using the number 2. We could consider various questions about 4^n. We might warm up with the familiar results. Where do we look for 4^2? – twice as far along as 4, and there indeed we find 16. Similarly we could do 4^3 and 4^4. What about $4^{3\frac{1}{2}}$? Going 3½ times as far as

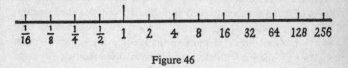

Figure 46

4 we find 128. We might then consider $4^{2\frac{1}{2}}$, $4^{1\frac{1}{2}}$, and $4^{\frac{1}{2}}$. What about 4^{-1}? That should be at -1 times the distance of 4, that is, the same distance but in the opposite direction. We find ¼ there.

This approach is particularly suitable, in that learners can become familiar with the logarithmic scale by using a slide rule. One can learn to work with a slide rule without knowing the theory of it, for instance by working through all the results in the multiplication table. One can observe how the powers of any number are distributed on the scale. One is then ready for the step to fractional and negative indices just described.

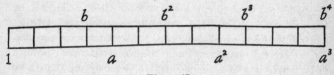

Figure 47

What shall we understand by an equation such as $b = a^{\frac{3}{4}}$? This means that *b* occurs ¾ as far along the scale as *a*. We can ask a learner to draw a diagram on squared paper to illustrate this situation, and to show the first few powers of *a* and *b*. The result will be as in Figure 47. He will observe that b^4 and a^3 coincide. And indeed this must happen. For b^4 occurs four times as far along as *b*, which is ¾ as far along as *a*. So b^4 must occur three times as far along as *a*, that is, at the same place as a^3.

Accordingly $b^4 = a^3$. By observing the roles that 3 and 4 play in this last equation and in the original question, a learner can soon realize how to convert an equation involving fractional indices into one that is free from them.

One particular case ought to be mentioned. What shall we understand by $b = a^0$? This is covered by our definition; a^0 is the number that occurs 0 times as far along the scale as a. It does not matter at what distance a occurs; 0 times that distance will give 0. So a^0 is the number marked at distance 0. Now on the logarithmic scale, distances are measured from the point where 1 is marked. Accordingly a^0 must mean 1, for every number a that appears on the logarithmic scale. You may wonder why the last part of the last sentence was included. The reason is that not all numbers appear on a logarithmic scale; negative numbers do not appear there, and there is no place for the number 0. Our definitions work for every number larger than 0. Certain fallacies can be produced by considering a symbol such as 0^0, which has no meaning. You may notice on the logarithmic scale that the numbers get smaller and smaller as we go to the left, but they never reach 0. If we thought of 0 as having a place on the scale at all, it would have to be an infinite distance to the left. To locate 0^0, we would have to multiply this infinite distance by 0, and this is something we just do not know how to do.

Our exercises on page 71 showed that a line of steepness 2 corresponded to an equation $y = ax^2$. With the freedom given us by fractional and negative indices, we can use a perfectly general rule; a line of steepness n on log-log paper corresponds to an equation $y = ax^n$. Here n may be any number, whole or fractional, positive or negative. If n is a simple fraction, we may use the device sketched above to put the equation in a form that will be understood by someone who is not at home with fractional indices. For instance, in our study of athletic records, we plotted s, the number of metres, against t, the number of seconds required to run that distance. The result on log-log paper was a straight line of slope 0·9, very nearly. This means that s is proportional to $t^{0·9}$, by the rule just mentioned. Now 0·9 is $\frac{9}{10}$. This leads to the result that s^{10} is proportional to t^9. (A little skill in algebra is needed here, to show that things work out just as simply when *proportional* is involved as they did earlier when a and b were related through an *equals* sign.) As $t = 10$ when $s = 100$, the record for the 100 metres being 10 seconds, we can find what the constant of proportionality must be. It turns out that $t^9 = s^{10}/10^{11}$ is a formula that gives, approximately, the world athletic records.

One reason why it is good to have some independent knowledge of these matters is that, with all the care in the world, books about mathematical topics are liable to contain misprints. For instance, John Clarke in his book *Population Geography* (Pergamon, 1965) discusses, among other things, the sizes of cities in relation to their rank. He quotes Lotka's rule for the hundred largest cities in U.S.A. as

$$(rank)\ 0.93 \times size = constant$$

Now clearly something is wrong here. In algebra, if three symbols are on the same level, as in *abc*, the understanding is that they are multiplied together. But nobody ever announces a law as rank times 0·93 times size is constant, for we could divide by 0·93 and reach the simpler law, rank times size is constant. (The constant would be a different one. For instance, if we write r for rank and s for size, the law in its first form might be $r(0.93)s = 1.86$, say. On dividing by 0·93, we would get $rs = 2$.)

Actually Lotka's rule is

$$(rank)^{0.93} \times size = constant.$$

It is in fact very much the same law as that we had earlier for the distribution of sizes of Canadian towns. (The term *rank* may be unfamiliar to some readers. It represents the quantity we called x when we were dealing with Canadian towns. To repeat our example, Ottawa being the sixth largest town has rank equal to 6.)

A certain familiarity with algebra thus helps one to detect misprints, and – with luck – to guess what the misprint should be replaced by. The consequences of accepting blindly a formula containing a misprint can be fatal.

THE ROLE OF LOGARITHMS

Logarithms have in the past usually been presented mainly as a way of performing calculations. Until recently, this was indeed their main function. However, with the spread of automatic calculating devices, large and small, this use of logarithms has become less important. If I had to multiply 2,379 by 6,183 I would sooner have a desk calculator than a table of logarithms. Perhaps logarithms are still useful for computing large powers of numbers, such as $(1·83)^{17}$. I did in fact use log tables, when writing the earlier part of this chapter, to find that at 2 per cent compound interest a population takes $55\frac{1}{2}$ years to treble. In the main, though, the present-day significance of logarithms lies in their use for stating or identifying a law (as in the work of this chapter) or for various theoretical arguments in which the logarithmic function occurs. The important thing is to understand the properties of logarithms. It is not necessary to sweat through a multitude of exercises in which logarithms are used – say – to solve triangles in trigonometry.

NOTE ON LOGIC

As was mentioned earlier, the seventeenth-century approach to logarithms, using repeated multiplication by $1·000001$, gave a procedure for constructing a logarithmic scale that was adequate for most practical purposes. However, it went by jumps and left gaps. To use our earlier illustration, it might show us where to mark 7 and $7·000007$ on the scale, but not show us where to put numbers between these, such as $7·000002$ say. In practice this causes little inconvenience, but the procedure is not congenial to a mathematician who wants to define log x for *every* number x. Certainly, this procedure has to be modified or elaborated or abandoned if we are to construct a rigorous theory of logarithms. But the construction of such a theory in any case belongs to a fairly advanced stage of mathematics. In the early stages of learning mathematics, the important thing is not absolute logic (which is rarely, if ever, attained) but acquiring a feeling for the subject and an ability to use mathematics correctly. Some years after acquiring this feeling for mathematics, it will be in order to come back and examine the logic more carefully and tighten up the definition of logarithm a little. But even at that stage, the account given here is not entirely without value, for much the neatest way to develop the theory is to use the order given in this chapter – first define the logarithmic scale, then use that to define x^k, where k may be fractional or negative or any number whatever.

CHAPTER FOUR

Some Notes on Algebra

FROM PICTURES TO FORMAL OPERATIONS

A YOUNG child, learning to work out $3 + 2$, may actually set out 3 objects and 2 objects and bring them together to make 5. Later on, learning to multiply 12 by 13, the child may use a picture such as Figure 105 in *Vision*. However, as the operations of addition and multiplication become more familiar, pictures and concrete aids gradually recede into the background of the mind. They should always remain as a resource available for coping with a problem or an uncertainty. But, as a rule, a person working out $2,375 \times 7,628$ by the traditional multiplication procedure neither expects nor needs to visualize the meaning of each step taken. In algebra too there are processes which we need to visualize at first but which later become more or less automatic.

If you put the following question to children, they will almost always solve it without using any special routine – 'I think of a number. I double it and add 3. The result is 7.' The children will work backwards. They will say, 'Before you added 3 you must have had 4. That was double your number, so you must have begun with 2.' In effect, here they are solving the equation $2x + 3 = 7$. If a teacher uses such a question in order to introduce the subject of solving equations, he finds himself in a quandary, because the children have solved the problem without any need for the special rules he is planning to teach them.

Now of course the reason for having set rules and procedures, in algebra as in arithmetic, is that we intend to handle questions much more complicated than the simple examples used to introduce ideas. For instance, an equation similar to $2x + 3 = 7$ may be met in the course of deriving a formula. Now the whole point of a formula is that it applies to situations that have not yet arisen; it must contain symbols each of which can be replaced by any number, so that it will apply to any problem (or a specific type) that may arise. Thus, instead of a particular equation $2x + 3 = 7$

we may meet the equation $ax + b = c$, where the numbers a, b, c depend on information we do not yet possess. Most learners, in my experience, do not give the solution of this equation with anything like the alacrity they show in solving $2x + 3 = 7$, in spite of the fact that the equations are of the same type and could indeed be handled by exactly the same argument. It therefore seems worth while to examine the solution of $2x + 3 = 7$ a little more closely and try to extract a general procedure from it.

Let us visualize this equation with the time-honoured device of the balance. Using a bag of marbles to represent the unknown x, the equation $2x + 3 = 7$ indicates 2 bags and 3 marbles balancing 7 marbles. The first statement in the children's natural solution – 'Before you added 3, you must have had 4' – corresponds to removing 3 marbles from each scale. That leaves 2 bags balancing 4 marbles. They then argue that each bag must contain $4 \div 2$, that is, 2 marbles.

Removing 3 marbles from each scale corresponds to the formal operation of subtraction, and leads to the very natural generalization – if the scales balance, they will still balance if you take away the same amount from each side. Equally, they would still balance if we added the same amount to both sides.

The operation $4 \div 2$ is one of division, and leads to the generalization – if the scales balance, they will still balance if you divide the amount on each side by the same number. This is very reasonable. If a cupful of sand weighs as much as a pint of milk, a quarter of a cupful of sand will balance a quarter of a pint of milk. Each quantity has been divided by 4. Equally, if it suited us, we could multiply and deduce that 5 cupfuls of sand would weigh as much as 5 pints of milk.

We can now solve the equation $ax + b = c$ by analogy with $2x + 3 = 7$, and state the formal steps as we go. We bear in mind that a plays the role of 2, b of 3, and c of 7.

Particular equation	General equation
$2x + 3 = 7$	$ax + b = c$
Subtract 3 from each side	Subtract b from each side
$2x = 4$	$ax = c - b$
Divide by 2	Divide by a
$x = 2$	$x = \dfrac{c - b}{a}$

Teachers usually insist that pupils state what steps they are taking, and the pupils usually think this is a waste of time and that the teachers are being fussy. However, there are two advantages in stating the steps. The first is that, if anyone else has to read the work, he can follow easily what is being done. The second is even more important. In real life you are not just working an exercise to be marked right or wrong. You are designing a building, or prescribing a dose for a patient, or staking your reputation on a scientific prediction. A small slip may mean that the building collapses, the patient dies, your reputation is destroyed. You therefore need to check your work several times. It is much easier to do this if you put your reasoning down on paper. Then you can come back later and check it just as you would the argument of another person. At some stage it says, 'Subtract a certain amount.' Did you in fact carry out this subtraction correctly? The real purpose of a formal calculation or a formal proof is not to convince anybody else; it is to convince you yourself that your reasoning has been sound. It is possible that pupils in schools would consider their teachers less fussy if mathematics were used more to plan something that was actually going to be made or done, and where the outcome was related to the real interests of the pupils.

FACT TEAMS

I have often noticed that children fail to see any connexion between the question, 'What do you have to add to 6 to get 10?' and the question, 'What is the result of subtracting 6 from 10?' This lack of understanding hinders them when they come to algebra, since they do not see any connexion between $x + 6 = 10$ and $x = 10 - 6$. My colleague Robert Wirtz has devised a procedure in the teaching of arithmetic to overcome this difficulty. He calls the four statements $6 + 4 = 10$, $4 + 6 = 10$, $4 = 10 - 6$ and $6 = 10 - 4$ a 'fact team'. In a certain sense, they all apply to the same situation. If I see a teacher of young children set out 6 buttons and 4 beads, I do not know if she is going to explain $6 + 4$ or $4 + 6$, or whether she is going to push the buttons aside and demonstrate that 6 taken from 10 leaves 4, or sweep the beads away and show that 4 from 10 leaves 6. The close relationship of the four facts is not a difficult one to perceive, and Wirtz gives exercises in which one fact, say $4 + 6 = 10$, is given, and the children are required to write the other three, $6 + 4 = 10$, $10 - 6 = 4$, $10 - 4 = 6$. In this way the association of the four facts becomes habitual. A child who had learnt arithmetic this way would find no difficulty in passing from $ax + b = c$ to $ax = c - b$; these would simply be two members of the same fact team.

In the same way, there are fact teams in multiplication and division, for example $2 \times 3 = 6$, $3 \times 2 = 6$, $2 = 6/3$, and $3 = 6/2$.

The equations $ax = c - b$ and $x = (c - b)/a$ can be regarded as two members of a fact team.

Familiarity with 'fact teams' in arithmetic should thus increase a learner's confidence when he comes to algebra.

One point of notation should be mentioned. The bracket in $(c - b)/a$ was necessary. A learner might easily forget this bracket and write $c - b/a$. With the values we had in our particular example, $c = 7$, $b = 3$, $a = 2$, $c - b/a$ would mean $7 - 3/2$, that is $7 - 1\frac{1}{2}$, or $5\frac{1}{2}$, which is certainly not the answer we want.

This difficulty could be avoided by using the form $\dfrac{c - b}{a}$ which is simple and convenient for someone working with pencil and paper. Unfortunately, it is very inconvenient for a printer. As much algebra reaches us through the printed page, it is well to be familiar with fractions in the form using the bracket. However, both when reading this book and when reading other printed material, you may find it helpful, as you work through some piece of algebra, to write the fractions in the form given at the beginning of this paragraph rather than in the bracket form.

TERMINOLOGY

In our work above, we began with the equation $ax + b = c$ and finished with $x = (c - b)/a$. The task we performed is sometimes described as 'solving the equation for x', sometimes as 'expressing x in terms of a, b, c'.

Of course there is nothing special about x. We could solve the equation for a; the result would be $a = (c - b)/x$. We could solve for b by finding $b = c - ax$. The equation is already in the form where c is given in terms of the other quantities; it is already solved for c. With the equation $ax + b = c$, the position is that if any three of the quantities a, b, c, x are given, the value of the fourth one follows. (We must exclude certain exceptional cases that lead to division by zero.)

Exercises

1. A family contains c children, of whom b are boys and g are girls.

   If there are 2 boys and 3 girls, how many children are there? What operation do you perform in answering this question? Write an equation giving c in terms of b and g.

   If there are 9 children, of whom 5 are boys, how many girls are there? What operation is used here? Write an equation giving g in terms of b and c.

2. Two numbers x and y have a total t. If t is 10 and y is 3 what is x? Solve the equation $x + y = t$ for x.

3. An article is normally sold at the price £p, but in a sale it is reduced by £r and offered at £s. If the normal price is £10 and the reduction £2 what is the sale price? Write an equation giving s in terms of p and r. Solve this equation for p. Also solve it for r. Check your answers by taking particular numbers for p, r, and s.

4. My gross income is £i a year but taxes take £t leaving an actual income £a a year. Write the Fact Team (four equations) corresponding to this situation.

5. Suppose that r rows with x objects in each row, give n objects in all.

   If there are 7 rows with 10 objects in each, how many objects in all? Write an equation giving n in terms of r and x.

   If there are 12 objects arranged in 3 rows, how many in each row? Solve your earlier equation for x.

   If there are 20 objects, with 4 in each row, how many rows? Solve your earlier equation for r.

6. A rectangle has base b, height h, and area A. The formula $A = bh$ is known to many people. Write the other members of the Fact Team (one equation involving multiplication, two involving division).

7. If V volts drive a current C ampères through a resistance of R ohms, then $V = CR$. Solve this equation (i) for C (ii) for R.

8. The momentum, p, of a mass m with velocity v, is given by $p = mv$. Solve (i) for v (ii) for m.

9. A batsman's average is given by $a = t/n$ where t is total runs scored and n is the number of times he has been out.

   If a batsman has an average of 20 runs and has been out 5 times, how many runs has he scored in all? Write an equation giving t in terms of a and n.

If a batsman has scored 120 runs and has an average of 30, how many times has he been out? Write an equation for n in terms of t and a.

10. The perimeter p (= distance around) of a triangle with sides a, b, c is given by $p = a + b + c$.

If $p = 9$, $b = 2$, $c = 4$ what is a? What operations do you perform in obtaining this answer? Write an equation for a in terms of p, b, c.

COMBINING PIECES OF INFORMATION

A typical use of algebra is to obtain a new result by combining several pieces of information. Frequently a textbook will start with several known scientific laws and show that some new law is a consequence of these. Sometimes certain data are expressed by equations and the system of equations is then solved; that is, it is shown that certain quantities must have certain values.

In Chapter 3 of *Vision in Elemenatry Mathematics*, we considered a problem about a man with twin sons. This problem was solved by drawing pictures, and then gradually the drawings were replaced by letters, until the solution appeared as conventional algebra.

The approach used in that chapter seems to correspond closely to the natural thinking of children. In various towns of the North American continent I have given demonstration lessons to children 9 or 10 years old – children with no previous experience of algebra and whom I had never met before. In every case, I have found the children capable, not merely of following the argument but of actively suggesting the steps to be taken, and reaching and understanding the algebraic treatment inside half-an-hour. With younger children, or with children below average intelligence, the same could probably be done provided more time was available.

Suppose someone works a number of problems of that type, with help from pictures whenever necessary, and has reached the stage where the pictures have gradually become superfluous and only algebra is being written down. We will consider what operations seem to be involved in the algebra.

Some Notes on Algebra

The problem of the man and his sons reached the form

$$m + s = 10 \qquad \text{(i)}$$

$$m + 2s = 14 \qquad \text{(ii)}$$

Then, following an idea suggested by the pictures, we compared these two equations. On the right, we saw that there was an increase of 4 as we went from 10 to 14. On the left hand, as we went from the first line to the second, an extra s appeared. This extra s, we concluded, must have been responsible for the extra 4, and we wrote $s = 4$.

What mathematical operation does *comparing* represent? When we say that 14 exceeds 10 by 4, we are carrying out subtraction; $4 = 14 - 10$. In the same way, when we say that $m + 2s$ contains an extra s as compared with $m + s$ we are subtracting $m + s$ from $m + 2s$. Accordingly a brief formal description of our argument might be that we have subtracted equation (i) from equation (ii).

It is not difficult to extract a general procedure from this example. In Figure 48, we see four blocks, the heights of which in feet are a, b, c, and d. The two tall blocks have the same height, and this is expressed by the equation $a = c$. The two short blocks have the same height, so $b = d$. We naturally conclude that $a - b$, which measures the height of the step on the left, equals $c - d$, which measures the step on the right.

Thus from the two equations

$$a = c$$
$$b = d$$

we can deduce the equation

$$a - b = c - d$$

It should be realized that in this scheme a, b, c, d may stand for fairly long expressions. For instance in the problem of the man and his sons, a was $m + 2s$ and b was $m + s$. This remark is akin to what was said on page 336 of *Vision*; at any stage of an arithmetical calculation we have simply a single number, but algebra gives us an indication of the steps by which we arrived at that number. Thus in Figure 48 the a-block might be required to be as high as the man and two sons, while the b-block might be required to represent the combined height of the man and one son. This would in no way harm the conclusion drawn from that figure, that if $a = c$ and $b = d$, then $a - c$ must equal $b - d$.

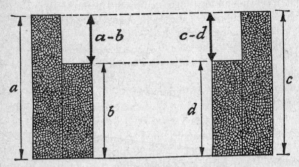

Figure 48

Subtraction is of course not the only operation that can be used in this way. Since the blocks on the left are exactly like the blocks on the right, putting the b-block on top of the a-block would give just the same height as putting the d-block on top of the c-block. The moral we draw is that if $a = c$ and $b = d$, then $a + b = c + d$. It is permissible to add equations.

Similarly, if we had a rectangular plot of sides a and b feet, it would be identical with a plot of sides c and d feet. The areas of these plots would be ab and cd, and of course these areas would be equal. So from $a = c$ and $b = d$ we can conclude $ab = cd$.

Arithmetic does not entirely prepare us for this kind of argument because in arithmetic we very rarely have occasion to say that two numbers are equal. If we are dealing with distinct numbers, such as 10 and 17, of course these are not equal. If the number 10 arises on two occasions, we think of this as the same number occurring again; we are not led to write $10 = 10$ and base an argument on this equation. Sometimes we perhaps verge on using the algebraic type of argument. For instance, to find 5 times 13 we might argue $13 = 10 + 3$, so 5 times 13 is 5 times $10 + 3$, and conclude that 5 times 13 is $50 + 15$, or 65. Even so, this sort of argument, using an equation, is not common enough or natural enough in arithmetic to establish a habit we can carry over to algebra.

One might put it this way; in arithmetic we do not use equations of the form $a = b$, because if a and b stand for the same number, the equality is obvious – it is a waste of time to write $10 = 10$; if they stand for different numbers, then they are not equal and the equation would be untrue.

To find arithmetical illustrations of our algebraic argument, we have to go away from the pure arithmetic of statements about numbers, such as $2 + 3 = 5$, and look for situations in the real world where the same number appears in two different roles. For instance, suppose a number of men and women attend some party and the host decides to give each guest a gift as a souvenir of the occasion. If it happened that the number of men was equal to the number of women, and the amount spent on each gift to a man was the same as the amount spent on each gift to a woman, we could conclude that the total amount spent on the men's gifts was the same as that spent on the women's. A multiplication of equations is involved here. The total expense on men's gifts is found by multiplying the number of men (m) by the value of a man's gift (a) and is thus ma (in suitable currency units). Similarly the total spent on the women's gifts is wb, where w women each receive a gift of value b. The equal number of men and women is expressed by the equation $m = w$; sex equality in the value of the gifts is shown by $a = b$; and the resulting equality of total expenditures by $ma = wb$.

It will be seen that the kind of argument we use in combining equations is an extremely reasonable one. If a learner finds difficulty at this stage, it is probably not because of any difficulty in the material itself, but because in years of previous work with arithmetic, the learner has formed an entirely wrong set of habits for approaching mathematics – looking for meaningless rules to memorize, instead of visualizing a situation and thinking about it.

There of course can be considerable difficulties, when we are confronted with a collection of equations, in deciding what sequence of arguments will lead us to a desired goal without impossibly heavy work along the way. This problem does not arise too often when we are reading mathematics, and simply have to follow the course selected by the author. Occasionally, of course, an author cheerfully skips half-a-dozen steps, and the reader then has, in effect, to solve the problem unaided. Such phrases as 'it is easy to see' or 'a brief calculation gives the result' should only be used by an author when the steps omitted are of a routine nature. In this book, I try to indicate some of the procedures which an author might reasonably assume to be routines that the reader will recognize and be able to supply.

When we are trying not simply to follow another but to attack a problem for ourselves, we are of course being much more ambitious. In real life, it is exceptional to be confronted by a single equation. Equations usually arise in batches. With even three or four equations, we are confronted by a very large number of choices as to the order in which they should be tackled and the operations that we should apply to them. We can only expect to weigh the situation up correctly and find an effective attack if we have had considerable practice and experience with this kind of work. Later in this book we shall meet many examples of equations as they arise in various branches of human knowledge. These will provide some indication of the methods that can be used and the considerations that guide us.

ON THE ABUNDANCE OF SOLUTIONS

In some circumstances a set of requirements can be met in very many ways; in other circumstances in only a few ways, or in none at all.

The same is true of a system of equations; there may be many solutions, or a few or one only, or none at all. It may seem foolish to consider equations with no solution, but such a situation can very well arise. When something is being designed, we wish it to perform well in a number of different ways. For example, in designing an aircraft, we wish the structure to be light so that the payload may be large. But we also wish the plane to be strong, so the parts under stress must be sturdy, and this drives the design in the direction of increased weight. Naturally it is possible to push these two conflicting requirements to the point where the project becomes impossible. If we are told to design a plane with a specified low weight and a specified high factor of safety, there may be no conceivable design that meets both requirements simultaneously. The corresponding equations have no solution.

At the other extreme, there are problems which permit of many solutions. In Conan Doyle's story *The Engineer's Thumb,* the engineer gets into a closed carriage in Eyford. He is then driven to a mysterious house, and estimates that the drive was about ten miles. Sherlock Holmes and Inspector Bradstreet agree that the mysterious house must lie on or within a circle of radius about ten miles drawn around Eyford. Clearly, there could be many mysterious houses meeting this condition.

In a design problem, if the specifications can be met in several different ways, the designer has a certain freedom of choice. In a scientific problem, we may seek further information, as Sherlock Holmes sought further ways of identifying the house. But the lack of equations may simply mean that nature permits something to happen in a variety of ways. If x carbon atoms combine with y hydrogen atoms so that all valencies are used, this can be expressed by suitable equations. But since very many different hydrocarbons exist, these equations cannot possible fix x and y uniquely. An abundance of solutions does not mean that something has gone wrong in our reasoning.

Graphs often provide a good method for deciding how many solutions are to be expected for a set of equations, and also help us to visualize what is happening. A very simple example is provided by our old problem of the man with twin sons.

Suppose we were given only the one equation $m + s = 10$. It is clear that this does not fix the heights of the man and the son. We can choose any two numbers that add up to 10. A solution such as $m = 7$, $s = 3$ can be shown graphically by plotting the point $(7, 3)$. Other solutions are $(6, 4)$ $(5, 5)$ and – if we are willing to ignore biological improbabilities – $(9, 1)$ $(2, 8)$, and a host of other points which, taken all together, fill the line AB in Figure 49. Any point that is on AB gives a solution of this single equation $m + s = 10$. No point that is off AB can satisfy this equation. Thus this equation, taken by itself, narrows down our search for a solution considerably but, like the engineer's recollections of his carriage ride, it does not fix a single solution, or even restrict us to some fixed number of possibilities. There are an infinite number of

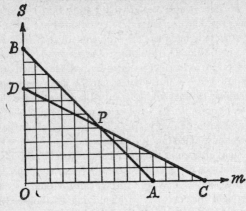

Figure 49

points on the line AB, and correspondingly the number of ways of satisfying this equation is infinite.

In the same way the solutions of $m + 2s = 14$ are infinite in number. They fill the line CD, where C and D correspond to the arithmetically convenient but practically unrealistic solutions $m = 14$, $s = 0$ and $m = 0$, $s = 7$. Any point that corresponds to a solution of this second equation must lie on the line CD.

Now the whole point of the puzzle is that both conditions must be satisfied. If we pick the point (8, 3), which does lie on CD but not on AB, even a young child will not accept the corresponding solution 'The man is 8 feet high and the son 3 feet.' The child will point out that these heights do not add up to 10 feet. Because the point is not on the first line, AB, the first condition is not realized.

The point that gives the solution must lie both on AB and on CD and it is clear that only one point does this – the point P with $m = 6$, $s = 4$, corresponding to the solution we obtained in Chapter 3 of *Vision*.

It can easily happen that a pair of equations have more than one solution. Suppose, for example, we have the equations $y = 2x + 12$ and $y = 10x - x^2$. The graph of the first equation is a straight line, of the second is an arch (see Figure 176, *Vision*, page 285.) The line meets the arch as shown in Figure 50, at the

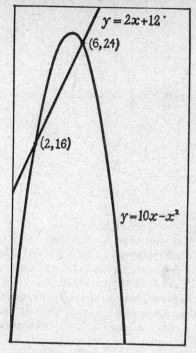

Figure 50

two points (2, 16) and (6, 24). It is easily verified that $x = 2, y = 16$ and $x = 6, y = 24$ satisfy both equations. Figure 50 makes it pretty clear that the line is not going to meet the arch anywhere else; we need not look for any more solutions.*

Another problem with two solutions is shown in Figure 51. O is the origin and A the point (6, 0). We want to find a point at a distance 5 from O and also at a distance 5 from A. Any point distant 5 from O must lie on the left-hand circle, and any point

---

*Strictly speaking, the figure shows that we should not expect any more *real* solutions. At a more advanced stage of mathematics, with important applications in electrical and mechanical work, we meet strange things called complex or imaginary solutions. These would not necessarily appear on a diagram of the kind used here.

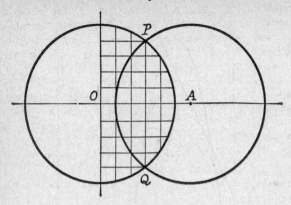

Figure 51

distant 5 from A must lie on the right-hand circle. These circles cross at the two points P and Q; P is (3, 4) and Q is (3, −4). As was mentioned in *Vision*, pages 279–81, circles are graphs of certain equations; the condition that a point is to lie on both of these circles could be expressed by writing a pair of equations, and this pair of equations would then have two solutions. The question of what kind of equation represents a circle will be taken up later in this book.

It will be noticed that when we have only one equation for two unknown quantities, the graph has been a line, as with $m + s = 10$, or a curve, as with $y = 10x − x^2$. In such situations there are infinitely many possible solutions. When we have two equations in two unknowns, we get two curves (it is convenient to regard a line as a special case of a curve for this purpose). The solutions occur where the two curves cross and this will normally happen only at a finite number of points.

As a rule then, if we are to fix two numbers, x and y, we need two equations. As 'fixing' we include situations like our last two examples, where there were a finite number of possible solutions.

The words *normally* and *as a rule* should be noted. The situation is affected both by the particular equations involved and by the kind of numbers we are prepared to accept for x and y. For instance, if x and y represented the number of men and women in a room, or the number of hydrogen and oxygen atoms in a chemical compound, so that only the counting numbers 0, 1, 2, 3, ... were acceptable values, the single equation $4x + 3y = 11$ would make it certain that $x = 2$ and $y = 1$. This is easily seen, since $4x$ is not to exceed 11, so x must be less than 3. If we try $x = 0$ or $x = 1$ we fail to get a whole number for y. Problems in which only whole number values are acceptable form a branch of mathematics known as the *Theory of Numbers*. Some problems in the theory of numbers are easy to state and yet so hard to solve that no one has yet managed to cope with them. The theory also contains a number of simple and interesting results, which, although they rarely have direct applications to other branches of knowledge, provide useful exercises in reasoning and the handling of algebra.

Usually, in practical problems, we are not dealing with whole numbers only but with what are known as the *real numbers*. In most of our graphical work we have been thinking of x and y as representing real numbers. When we are plotting a point, we do not object if x and y lie between the whole numbers, as for example the point $(1\frac{1}{2}, 2\frac{1}{4})$. We do not object to the presence of minus signs; $(-3, 5)$ is a perfectly good point. By allowing x to have either a $+$ or $-$ sign, we ensure that we can move either to the right or the left. Further, we can go any distance we like. We are not restricted to whole numbers like children who have just learned to count; corresponding to any length whatever, there is a real number that measures it.

Even with real numbers, it can happen that two quantities are fixed by one equation. Consider the equation $x^2 + y^2 = 0$. In Chapter 12 of *Vision* we were led by several different considerations to the result 'minus times minus is plus'. This means that if we multiply -3 by -3 we get $+9$, and the same happens with any other negative number; its square is positive. Naturally the square of a positive number is positive. The square of 0 is 0; for any real number except zero, the square is positive, that is, greater than 0. Now if we add two positive numbers together, the result is bound to be greater than 0. So if x^2 is positive and y^2 is positive, the sum $x^2 + y^2$ will exceed 0, and we shall not have a solution of $x^2 + y^2 = 0$. The only escape is to take both x and y to be 0. So the only solution of $x^2 + y^2 = 0$ is $x = 0$, $y = 0$.

One way to look at this is the following. We shall show later that the circle with centre at the origin and radius r is $x^2 + y^2 = r^2$. If we put $r = 0$ we get $x^2 + y^2 = 0$, so this equation can be thought of as representing a circle of radius 0, which consists of a single point.

When we have a system of several equations, there are various peculiarities that can arise, which have not been discussed here. It is therefore not feasible for me to state anything in the nature of a precise, mathematical theorem. However there is something which may be called a useful doctrine. Generally speaking, we need two equations to fix two quantities x, y. 'Generally speaking' sounds very vague. What it means in practice is this. Suppose I have to deal with some situation in physics. In looking at the situation, I see that certain physical laws apply. Each law allows me to write down an equation. This is the first phase of dealing with the problem – to analyse the situation and write down equations. The second phase is to solve the equations. How shall I know when I have written down enough equations to complete the first phase? If the situation involves two unknowns, my practice is to write down two equations and try to solve them. In trying to solve them, I may run into some kind of difficulty or exceptional state of affairs, which I shall then recognize. But until that happens, I work in the expectation that two equations will be enough to fix my two unknowns. In the same way, if there were three unknowns, I would not embark on trying to determine them until I had three equations, and generally if there were n unknowns, I would assemble n equations before trying to find the solution or solutions.

SUBSTITUTING AND ELIMINATING

In our problem of the man with two sons there was a stage where we had found $s = 4$. We then turned to the equation $m + s = 10$ and argued, 'If s is 4, this equation means that $m + 4 = 10$.' The process used here is formally referred to as *substitution*. We have found that s equals something, so we feel justified in replacing s by that something.

The 'something' need not always be a single number. For instance, we might be dealing with a system of equations that included $x = y + z$ and $x + 2y + 3z = 11$. The first equation tells us that we may replace x by $y + z$. When we do this in the second equation, we obtain $(y + z) + 2y + 3z = 11$, which simplifies to $3y + 4z = 11$.

The reason why a system of equations does not immediately tell us the values of the quantities involved is that each equation is liable to refer to several unknowns. For instance, in the problem of the man with two sons, the equation $m + s = 10$ involves the heights both of the man and a son, so that it does not fix either height. We solved that problem by manoeuvring until we obtained an equation, $s = 4$, that involved just the one unknown, s.

Substitution often gives us a way of reaching our goal. Instead of the method used earlier to solve the problem of the man and the sons, we could use a method based on substitution. From the first equation, $m + s = 10$, we would find $m = 10 - s$. This gives us the 'something' to replace m in the second equation, $m + 2s = 14$. Replacing m by $10 - s$ changes the second equation to $(10 - s) + 2s = 14$, which means $10 + s = 14$ and so $s = 4$.

We could describe the strategy just followed thus – solve the first equation for m and substitute the result in the second equation.

The substitution procedure will often give us a systematic way of solving a system of several equations. The danger, when you have to deal with several equations, is that you will wander about without any sense of direction and reach no solution. The great thing therefore is to see your strategy as a whole, before you get down to detailed calculations.

The plan to be followed may be seen from the following example. Suppose we have to deal with the equations

$$
\begin{aligned}
x + y + z &= 7 &\text{(i)}\\
x + 2y + 3z &= 17 &\text{(ii)}\\
x + 3y + 2z &= 15 &\text{(iii)}
\end{aligned}
$$

We can solve the first equation for x and find $x = 7 - y - z$. If we now replace x by $7 - y - z$ in both the equations (ii) and (iii), we shall obtain two equations in which only y and z appear. We are thus nearer our goal. We began with three equations in three unknowns; now we have two equations in two unknowns. These equations would in fact be $(7 - y - z) + 2y + 3z = 17$

and $(7 - y - z) + 3y + 2z = 15$. These simplify to the following

$$y + 2z = 10 \qquad \text{(iv)}$$
$$2y + z = 8 \qquad \text{(v)}$$

We could, if we liked, solve these by the kind of method considered in Chapter III of *Vision*. Or we could continue to use substitution, and this perhaps brings out the underlying, systematic idea better. We have succeeded in getting rid of x; we go on to get rid of y. Solving (iv) for y gives us $y = 10 - 2z$. That means $2y = 20 - 4z$. (Our reason for making this calculation is that equation (v) contains not simply y but rather $2y$.) Replacing $2y$ by $20 - 4z$ in equation (v) gives $(20 - 4z) + z = 8$ or $20 - 3z = 8$, and this tells us $z = 4$. A few lines earlier we had the result $y = 10 - 2z$. Now that we know $z = 4$ we see $y = 2$. Our first step of all gave us $x = 7 - y - z$. We know now what y and z are, so we see $x = 7 - 2 - 4 = 1$.

The strategy of successive substitution used here can be applied to any number of equations. For example, if we had five equations involving v, w, x, y, z, we might solve the first equation for v. On substituting for v in all the other four equations, we would obtain four equations containing only w, x, y, z. We then apply the same idea again. We might solve the first of these equations for w in terms of x, y, z. Substituting for w in the other equations then gives three equations in x, y, z. We continue along the same lines, and end up with a single equation for z.

Of course there is no obligation to solve for v, w, x, y, z in their alphabetical order. If the first equation were, say, $7v - 2w + 5x + y + 3z = 11$, it would be much easier arithmetically to solve for y and then, by substituting in the other equations, find four equations involving v, w, x, z.

The process of getting a system of equations in which x does not appear is often called *eliminating* x.

It will be seen that in the course of solving a system of equations, we are frequently obliged to make small, routine calculations of the type discussed in *Vision*. For instance, in solving the system (i), (ii), (iii) above we had to double $10 - 2z$. This gave $20 - 4z$, to which we then had to add z. If difficulties are found in some detail of this work, it is possible that the discussions of algebraic computation in Chapters 3, 9, and 10 in *Vision* may resolve the difficulty.

Beginners in algebra may find a number of small difficulties as they go from particular cases, involving numbers, to general cases involving algebraic symbols. This may be seen by comparing the following two worked exercises. The first is straightforward. The second follows essentially the same pattern, but there is one point at which a beginner may stick.

Example 1. Solve the system of equations $x = 3z$, $y = 2z$, $x + y + z = 60$. From the first two equations we see that x may be replaced by $3z$ and y by $2z$. Making these replacements in the last equation, we find $3z + 2z + z = 60$, that is $6z = 60$. So $z = 10$ and accordingly $x = 30$, $y = 20$.

Example 2. Solve the system of equations $x = az$, $y = bz$, $x + y + z = c$, where a, b, c represent known numbers. The first two equations show that x may be replaced by az and y by bz. The last equation thus gives $az + bz + z = c$.

It is at this point that a learner may be uncertain what to do next. If we look back at Example 1, which provides a model for what we are doing now, we see the next step is $6z = 60$. Where did the 6 come from? The 6 shows the total number of z's on the left-hand side. It came from the addition $3 + 2 + 1$. In our present example, 3 has been replaced by any number a and 2 by any number b. The single z at the end remains unaltered. So $3 + 2 + 1$ in the particular case needs to be replaced by $a + b + 1$ in the general case. One might be tempted to replace $6z$ by $a + b + 1z$, but this would mean that z was multiplied only by 1. In the particular case, z is multiplied by the whole of 6; in arithmetical work we take this for granted and never think of doing anything else. But in algebra we have not merely a number like 6 but a history of how that number came to be there, so that we see pieces inside it which were invisible in the arithmetic. Accordingly we have to emphasize that we are multiplying z by the whole of this number. Our next step accordingly is written $(a + b + 1)z = c$. Finally we solve for z. We may think in terms of Fact Teams, or we may divide both sides of the equation by $a + b + 1$. Either way we reach $z = c/(a + b + 1)$. Since $x = az$, x must be a times this, that is $x = ac/(a + b + 1)$ and similarly, from $y = bz$, we have $y = bc/(a + b + 1)$. (Fractions did not appear in Example 1,

owing to the particular choice of the numbers in it. If any difficulty is found with the last steps here, it may be helpful to consider the particular values $a = 4, b = 2, c = 5$. This should make
clear what operations are being performed on the fractions and
show that these operations are justified.)

Exercises

Solve the systems of equations below.

1. $x = y, x + y = 10$.
2. $x = y, 2x + y = 12$.
3. $x = y + 1, x + y = 21$.
4. $x = y + 1, 2x + y = 14$.
5. $x = y + 10, 3x + y = 34$.
6. $x = 10 - y, x + 3y = 14$.
7. $x = 10 - y, x + 4y = 19$.
8. $x = 10 - y, x + 2y = 10 + z, y + z = 8$.
9. $x + y = 10, x + 2y = 11 + z, y + z = 11$.
10. $x + y = 10, x + 2y + z = 13, x + 3y + 3z = 18$.
11. $x = y, x + y = 1$.
12. $x = 2y, x + y = 1$.
13. $x = 3y, x + y = 1$.
14. $x = ay, x + y = 1$ where the number a is supposed known.

In questions 15 to 18, a and b represent known numbers.

15. $x = ay, x + y = b$.
16. $x = y + a, x + y = b$.
17. $x = ay, x + by = 1$.
18. $x = ay, bx + y = 1$.

A SMALL ROUTINE

In Example 2 of the previous section we discussed at some
length that $az + bz + z$ could be replaced by $(a + b + 1)z$. A
step of this kind occurs so frequently as a small detail in a long
calculation that most users of algebra come to regard it as routine.
At a certain stage of learning algebra this is perfectly justified. A
beginner however should always go slowly and make sure that
each step he takes is a sensible one. For instance, he can see that
the step mentioned above is reasonable by using the kind of
diagram explained in Chapters 9 and 10 of *Vision*. He should
consider the step in the opposite direction. If he had to multiply
out $(a + b + 1)z$ he might draw the simple diagram of Figure 52,

and obtain the result $az + bz + z$. Both expressions, in the language of Chapter 9 of *Vision*, represent the 'number of dots' imagined in Figure 52, and accordingly must be equal.

In Figure 52, three rectangles join together to make one large rectangle. The small rectangles contain z rows of a dots, z rows of b dots, and z rows of 1 dot. They fit together to make one large rectangle because they each have the same number of rows. Someone starting from the expression $az + bz + z$ and wondering if it could be dealt with by a diagram like Figure 52, might see that it could be, because the letter z occurs in each term of the expression. We see z in az, in bz, and in z, so each of these can be pictured by means of a rectangle containing z rows.

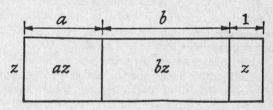

Figure 52

After working with algebra for a while, it becomes both tedious and unnecessary to draw pictures and think through the whole business from the beginning. After doing many multiplications in algebra, one comes to know that $(p + q + 5)x$ will be $px + qx + 5x$, and, when going in the other direction, one sees x in each term of the second expression and immediately knows that it can be replaced by the first.

Anyone who works fairly regularly with algebra will find this transition occurs quite naturally and gradually. Any time you are in doubt you can go back to the picture and the detailed argument, and perhaps check your work by trying a few particular numbers for the algebraic symbols a, b, p, q, x, z. But gradually, you will find yourself more and more looking at the actual letters and knowing what you can do with them, and less and less going back to the arithmetic they represent. This is perfectly healthy and natural at a certain stage. It is disastrous if someone who has reached this stage thinks he can teach a beginner to 'do' algebra by applying rules to the letters. A beginner must begin by thinking what the letters are saying about arithmetic. And even an expert at algebraic manipulation needs sometimes to remember that algebra had its origin in arithmetic.

The operation just described often has to be applied to more than one letter in an algebraic expression. For instance, if in the course of a calculation we met the expression $4x + 5y + 3x - 2y$ we would naturally reflect that, since the order in which additions and subtractions are carried out does not affect the final answer, this could be written as $4x + 3x + 5y - 2y$, which would then boil down to $7x + 3y$. In a more general calculation we might have to deal with $ax + by + cx - dy$, in which the symbols a, b, c, d have replaced the particular numbers 4, 5, 3, 2. The same steps that we used before would lead us to replace this expression first by $ax + cx + by - dy$ and then by $(a + c)x + (b - d)y$.

Later in this book Chapter 8, *A Method of Discovery*, presents, in the guise of a trick, something which is actually a very powerful method for finding the law behind a series of observations. In that chapter the 'conjuror' says, 'The law behind the numbers you gave me is $y = 3x^2 + 9x + 11$' or something like that. Then the question is posed, 'How did you know to put $9x$ in the middle of your formula?' The conjuror answers, say, 'To get that 9 I subtracted *this* number from *that* one' and he points to two numbers written on the board. If we call the numbers at which he has pointed h and k, he has found 9, the coefficient of x, by working out $k - h$. If we replaced 9 by $k - h$, the formula would become $y = 3x^2 + (k - h)x + 11$, which shows now not merely the number, 9, that he got for the particular sequence of numbers he has been given, but the rule he would use to find the number associated with x for any similar sequence. The question of course would arise – how did he know to start his formula with $3x^2$ and to end it with 11? If he answered these questions, his formula would take the shape $y = (\ldots)x^2 + (k - h)x + (\ldots)$ where the brackets would contain expressions showing the rules by which he arrived at the coefficient of x^2 and the constant term. This is mentioned to show that learning how to deal with expressions such as $(a + b + 1)z$ or $(p + q + 5)x$ is not a useless exercise. Such expressions do turn up in quite practical questions.

Readers should not search Chapter 8 to find where the conjuror gives the law $3x^2 + 9x + 11$ or the rule $k - h$ for the coefficient of x. Chapter 8 is meant to be read as an experience in discovery, and I do not want at this stage to give away any details of how the conjuror does his trick. These details will all be found in Chapter 8 in their proper places when the chapter is read as a whole. The present remarks are meant purely to indicate the kind of thing that goes on in that chapter. and that our present routine work does have significant applications. I do not think the formula $y = 3x^2 + 9x + 11$ or the expression $k - h$ actually occurs anywhere in Chapter 8.

A PROBLEM FROM CHEMISTRY

Systems of equations arise naturally in chemistry. These systems are of interest to people who are not chemists, as they make good examples of problems involving several very simple equations; also, you do not need to know any chemistry to understand them.

Atoms in chemistry are like individuals in society. They are represented by symbols, such as H for a hydrogen atom, O for an oxygen atom, Cl for a chlorine atom. Atoms, like human individuals, can exist in solitude but they mostly prefer not to. Hydrogen atoms are normally found in pairs, called hydrogen molecules, H_2, and oxygen also in pairs O_2. Many others groupings can occur. Almost everyone today knows that the water molecule is H_2O, a stable social unit consisting of two hydrogen atoms and one oxygen atom. A chemical reaction is a change from one social grouping to another, and is shown by an equation, which indicates what groups enter and what groups emerge.

Knowing that some H_2 groups and some O_2 groups can go into a state of flux and emerge as a number of H_2O groups, we naturally want to know what the numbers in question are. If x of the H_2 combine with y of the O_2 to make z of the H_2O, what numbers should x, y, and z be?

As mentioned above, no knowledge of chemistry is needed to deal with this question. You can think, if you like, of hydrogen molecules, H_2, as packets containing two red balls, and oxygen molecules O_2, as packets containing two black balls. We plan to open some packets and make new packets each containing two red balls and one black ball, H_2O.

The process would be represented by a chemical equation $x\mathrm{H}_2 + y\mathrm{O}_2 \rightarrow z\mathrm{H}_2\mathrm{O}$. The only information we have is that the number of H does not change (no red balls are lost or introduced from outside) and similarly the number of O does not change. Each of these statements gives us an equation.

The number of H atoms at the start of the process is $2x$, since there are two of them in each of the x packets. At the end of the process, there are $2z$ of the H atoms, since there are two of them in each of z packets. As the number does not change, $2x = 2z$, which means $x = z$.

We deal similarly with the O atoms. We begin with $2y$ of them. We end with z of them, since there is only a single O in each of the z packets. Accordingly, $2y = z$.

We conclude that z must be twice as large as y and that x must be the same as z. This is all the information we have or can obtain. Accordingly there are many solutions; we can choose any number we like for y. For instance if we took $y = 10$, we should find both x and z to be 20. Equally well, we could take y to be a million, and each of x and z to be two million. In fact, if we were manufacturing water from hydrogen and oxygen, we could make it on any scale we chose; we could make a few drops or a few gallons. There would definitely be something wrong if our equations fixed x, y, and z completely.

In chemistry, it is usual to choose the smallest solution in whole numbers. We would choose $y = 1$ and get $x = 2$, $z = 2$ and the usual equation $2\mathrm{H}_2 + \mathrm{O}_2 = 2\mathrm{H}_2\mathrm{O}$.

However, if we had not chemistry in mind and were asked to solve the system of equations $x = z$, $z = 2y$, we would have to give the most general solution, covering all possibilities. Since y can be chosen equal to any number, we might write n for any number and give the solution in the form $x = 2n$, $y = n$, $z = 2n$, where n may be chosen at will. It would depend on the circumstances whether n had to be chosen from the whole numbers, from the real numbers, or perhaps from some other set of acceptable numbers.

In the system $x = z$, $z = 2y$ we have only two equations for the three unknowns x, y, z. As we discussed earlier, usually three equations are required to fix three unknowns, so it is not surprising that our two equations leave us some freedom. When we choose y, so that it ceases to be unknown, we are left with two unknowns x and z, and the two equations are sufficient to fix these in terms of y.

A GLUT OF EQUATIONS

Counting equations however does not always tell us how things are going to work out. There is a reaction between $Al_2S_3O_{12}$ and CaH_2O_2 which produces $CaSO_4$ and AlH_3O_3, so we have a process $xAl_2S_3O_{12} + yCaH_2O_2 \rightarrow zCaSO_4 + tAlH_3O_3$. Here we have only four unknowns, x, y, z, t. But we have five elements, Al (aluminium), Ca (calcium), H (hydrogen), O (oxygen), and S (sulphur). There will be an equation for each of these elements, to say that the number of its atoms stays unchanged. So we have five equations for four unknowns, for which it is usually impossible to find any solution at all, let alone enough solutions to allow the reaction to occur on any scale we choose. However the chemistry textbooks assure us that this reaction does take place, so there must be something special about the five equations, which we shall recognize when we are solving them. The five equations are listed below; opposite each equation is written the chemical element, whose atoms it shows to be unchanged in number.

Al	$2x = t$
Ca	$y = z$
S	$3x = z$
H	$2y = 3t$
O	$12x + 2y = 4z + 3t$

The first three equations show that, if we choose x, the values of y, z, and t are fixed; $t = 2x$, $z = 3x$, and, from the second equation, y must be the same as z, so $y = 3x$ also. Accordingly, in the last two equations, we can substitute for y, z, and t, and get equations involving x alone. But when we do this, we get no information. Replacing y by $3x$ and t by $2x$ in the H equation we simply get $6x = 6x$. Substituting in the O equation gives us $12x + 6x = 12x + 6x$. Neither of these helps us to fix x, since both equations are automatically true for every number that exists. Thus our choice of x is still completely free, and we have in fact the situation we expect in a chemical problem; the proportions in which the various elements enter are fixed, but there is nothing to tell us the actual numbers of atoms involved. For the purpose of writing the chemical equation, we would select the smallest solution, $x = 1$, $y = 3$, $z = 3$, $t = 2$.

It is to be noticed that we have just solved a system of five equations, which sounds rather terrifying. However, by picking one's way carefully, such systems are often not hard to handle. The equations were not dealt with in any plan based on alphabetical order, but in an order that gave the simplest working. The oxygen equation, which is much longer than any of the others, was left till last.

A point that arises in some chemical equations may be illustrated by the reaction $xKClO_3 \rightarrow yKCl + zO_2$. This leads to the equations $x = y$, $3x = 2z$. The first of these leads to no difficulty, but in the second equation neither x nor z is a whole number times the other. Whether we solve for x or z we find fractions coming in, which is not very convenient when we are looking for whole number solutions. The difficulty may be met in the following way. Since $3x$ is $2z$, that means $3x$ is an even number, so x itself must be even. So, for some n, $x = 2n$ (*Vision*, p. 15). Substituting $2n$ for x in the equation $3x = 2z$, we find $6n = 2z$, which means $z = 3n$. Accordingly we have $x = 2n$, $z = 3n$, and we can substitute for both x and z in terms of this new symbol, n, without introducing fractions into the work. The complete solution would be $x = 2n$, $y = 2n$, $z = 3n$ and the smallest solution would be found by taking $n = 1$, namely $x = 2$, $y = 2$, $z = 3$.

Exercises

Find the smallest solution required for writing the chemical equation in the following cases:

1. The rusting of iron: $xFe + yO_2 \rightarrow zFe_2O_3$
2. $xZn + yHCl \rightarrow zZnCl_2 + tH_2$
3. $xFeCl_3 + yNaOH \rightarrow zFeO_3H_3 + tNaCl$
4. $xCa + yH_3PO_4 \rightarrow zCa_3P_2O_8 + tH_2$
5. $xN_2H_4O_3 \rightarrow yN_2O + zH_2O$
6. $xCaO + yH_3PO_4 \rightarrow zCa_3P_2O_8 + tH_2O$
7. The burning of benzene: $xC_6H_6 + yO_2 \rightarrow zCO_2 + tH_2O$
8. The body's reaction to alcohol:
$$xC_2H_6O + yO_2 \rightarrow zC_2H_4O_2 + tH_2O$$

The reaction actually occurs in two stages (see p. 61 of Roueché, *Alcohol*, Grove Press, New York).

9. $x\text{Cu} + y\text{HNO}_3 \rightarrow z\text{CuN}_2\text{O}_6 + t\text{NO}_2 + u\text{H}_2\text{O}$

10. The explosion of nitroglycerine:

$$x\text{C}_3\text{H}_5\text{N}_3\text{O}_9 \rightarrow y\text{CO}_2 + z\text{H}_2\text{O} + t\text{N}_2 + u\text{O}_2$$

CHAPTER FIVE

Some Questions of Notation

NOTHING in this chapter involves any new mathematical idea or result. The chapter is essentially concerned with a number of abbreviations, signs that save us the trouble of writing phrases in English or long mathematical expressions. Yet it is well worth including, as experience shows that readers of mathematics will sometimes be put right off by meeting a number of unfamiliar signs. The appearance of these can convey the suggestion that the work is extremely mysterious and difficult, when in fact it is merely some fairly simple statement in unfamiliar dress.

LESS THAN AND MORE THAN

The sign $>$ means 'greater than', 'larger than' or 'more than'; the sign $<$ means 'smaller than' or 'less than'. Whoever invented the signs $=, >, <$ may well have been led to do so by envisaging situations such as those shown in Figure 53. The signs reproduce the directions of the dotted lines which emphasize the relationships *as big as, bigger than, smaller than*. These pictures may help one to remember which of the signs $>$ and $<$ means 'larger than' and which 'smaller than'.

These signs are sometimes combined. For instance, the numbers on dice go from 1 to 6. If x stands for a whole number that may be found on dice, we may wish to express the fact that x can be as much as 6 but not more than 6. The situation here is that x may either be 6 or a number less than 6. In signs either $x = 6$ or $x < 6$. To save the words 'either . . . or' the signs are often written one above the other, like this $x \leqq 6$, and this is very often compressed to $x \leqslant 6$. Either of these would be read as 'x is less than or equal to 6'. In the same way $x \geqslant 1$ expresses the other fact we noted about the numbers on dice, and is read as 'x is greater than or equal to 1'.

In this book such signs do not play any role, but in both mathematical and scientific literature they are liable to occur. The reason

The man is taller than the boy

The girl is shorter than the woman

Greater than

Less than

The boy is just as tall as the girl

Equal to

Figure 53

for this is that we are rarely able to cope with the full complexity of a situation either in the real world or in a mathematical problem. We frequently try to identify the main forces at work and to calculate their effect; we have to establish that our conclusion has not been vitiated by the factors we have neglected, and to do this, we do not discuss their effects in detail, but try to establish that these effects will be less than a certain amount.

In working with inequalities (that is, relations of *more* or *less*) it is important always to think realistically and not to be carried away by formal arguments that sound all right but in fact lead to false conclusions. With equations we get used to the idea that we may 'do the same thing' to both sides of an equation and thus arrive at another true equation. One has to be very wary of such arguments with the signs $>$ and $<$. For instance, it is true that $-3 < -1$. (Someone in debt for £3 is *less* well off than someone in debt for £1.) Someone might suggest, 'Square both sides.' Now $(-3)^2 = 9$ and $(-1)^2 = 1$, so the result would be $9 < 1$, which is certainly false; 9 is not less than 1. Even where negative numbers are excluded, one has to be careful. For instance it is true that $2 < 7$. Someone might suggest, 'Subtract both sides from 10.' This would work all right with an equation, but here it would produce the untrue statement $8 < 3$.

Sometimes a crossing-out line is used to express 'not'. Thus $\neq$ would mean 'is not equal to'. For example, we can divide by any number except 0. Thus, if we plan to divide an equation by x we have to insert 'provided x is not 0', which may be put in the compact form '$x \neq 0$'.

SUBSCRIPTS

Another device which is essentially simple but can look unfamiliar and difficult is that of subscripts. We met these, very briefly, when we were considering the survival of mountain sheep, in the chapter 'Mathematics through the Eye'. Here we will consider another example of survival in nature. Lack, in his book *The Life of the Robin* (Witherby, 1943) gave some tables to show how long robins live. He observed 144 robins who had just left the nest and saw how many were alive on 1 June of each succeeding year. His table read as follows:

Year	Alive on June 1	Died in following 12 months
0	144	111
1	33	15
2	18	14
3	4	2
4	2	1
5	1	1
6	0	0

If we want to refer to some figure in this table, we can explain without difficulty which figure we have in mind. For instance, 18 is the number of robins alive on their second birthday; 14 is the number that die between their second and third birthdays (we regard 1 June as the 'birthday' of all robins). Someone doing research on a great number of such tables might well wish for a shorter form of reference and devise the following scheme.

Year	Alive on June 1	Died in following 12 months
0	L_0	D_0
1	L_1	D_1
2	L_2	D_2
3	L_3	D_3
4	L_4	D_4
5	L_5	D_5
6	L_6	D_6

Here L_4 stands for the number still *living* on their fourth birthday, D_4 for the number that *die* at the age of four.

Some learners are disturbed by the fact that a single number is described by a symbol such as L_4 which contains two parts, L and 4. We could use single letters – say a robins alive at the start, b still alive a year later, c still alive 2 years later and so on. We could use p for the number dying in their first year, q for the number dying in their next year, and so on. But each time we used one of the symbols a, b, c, . . ., p, q, r, . . . we would have to pause and think what it meant. On the other hand, with the subscript notation used above, on seeing L_4 we immediately know what it indicates; L stands for *living* and 4 indicates 'on the fourth birthday'.

None of Lack's robins reached their sixth birthday. Now it might happen in some such investigation that an exceptionally hardy or cunning robin reached its seventh birthday. If we had used single letters, a, b, c, d, e, f, g to indicate the numbers surviving to birthdays 0, 1, 2, 3, 4, 5, 6, we would have to define a new symbol, h, for the number reaching birthday 7 and indicate the single survivor by writing $h = 1$. With the subscript notation we do not have to explain any new symbol if a creature lives unusually long; the equation $L_7 = 1$ is understood without further explanation.

The single letter system is particularly inconvenient with long-lived creatures such as human beings. In Canada* one woman in every 100,000 reaches the age of 110. It would be most inconvenient to have a list of symbols covering every age from 0 to 110. The subscript notation provides us with a ready-made scheme; L_{83} for instance represents the number surviving to their eighty-third birthday.

A good way to become accustomed to the subscript notation is to take some statement which you understand and to translate it into the subscript notation. The exercises below are based on this idea. We start with Lack's table, which I imagine everyone will find quite straightforward, and try to express that information through subscript abbreviations.

Exercises

1. What L or D symbols correspond to the following numbers in Lack's table?
    (i) 33, (ii) 4, (iii) 14, (iv) 15, (v) 111, (vi) 144.
2. In Lack's table, what are the values of
    (i) L_2, (ii) D_2, (iii) L_3?
3. Given that 18 robins reach their second birthday and that 14 die in the following year, we can work out that 4 will be alive on their third birthday.
    (i) What arithmetical operation do we use in this calculation?
    (ii) What equation shows how 4 comes from the given numbers, 18 and 14? (iii) What equation results, if we replace 18, 14, and 4 by the corresponding L and D symbols?
4. (i) If in some survival table you knew the values of L_1 and D_1, would it be possible to calculate the value of L_2? (ii) What equation would give L_2?
5. If in some survival table the values of L_n and D_n were known, of what other symbol could the value be deduced? (ii) What equation would give it?
6. (i) 33 robins reached their first birthday and 18 their second. What other number in Lack's table could be deduced from this information?
    (ii) Write the calculation just made in the form of an equation $33 \ldots 18 = \ldots$ (The dots indicate that a sign is missing on the left hand and a number on the right.)
    (iii) In your answer to (ii), replace the numbers by the corresponding L and D symbols.

*See A. Pedoe, *Life Insurance, Annuities and Pensions*, University of Toronto, 1964.

(iv) Write an equation showing how the value of D_n can be found from the values of two L symbols.

That is, find a generalization of your answer to (iii).

(v) Is your answer to (iv) a guess, or can you be certain, by considering the meanings of the symbols, that it is correct?

7. (i) We suppose you are provided only with the information in the 'Died' column of Lack's table. From this, determine how many robins died before their third birthday.

(ii) Express in terms of the D symbols the calculation you made to answer (i).

(iii) How would you find the number of dying before their third birthday if you had available only the figures in the 'Alive' column?

(iv) Write an expression, using only L symbols, to show what you calculated when you answered (iii).

(v) You found formulas in parts (ii) and (iv); both formulas were for finding the same thing. Write an equation that expresses the fact that these two formulas must always give the same number.

8. Answer the same questions as in question 7 above, but replace 'third birthday' by 'fifth birthday'.

Questions 7 and 8 are intended as preparation for the ideas of the next section; it is desirable to answer them before reading on.

THE SUMMATION SYMBOL

In question 7(i) of the last section it was required to find how many robins died before their third birthday. Lack's table shows that 111 robins died in their first year of life, 15 died at age 1, and 14 at age 2. The total that died aged less than 3 is thus $111 + 15 + 14$. The method is perfectly general; if in some other experiment we needed to know the number dying before age 3, we would add together the numbers dying at ages 0, 1, and 2. In symbols this would appear as $D_0 + D_1 + D_2$.

Similarly, question 8 led to the conclusion that the number dying aged less than 5 was found by adding together the numbers dying at ages 0, 1, 2, 3, and 4; in symbols, the number is given by $D_0 + D_1 + D_2 + D_3 + D_4$.

If for a human population we wanted to find the number dying under the age 50 we would have to do an even longer addition; the required number would be $D_0 + D_1 + D_2 + \ldots + D_{48} + D_{49}$. Obviously it would be very tedious to write out such a long expression in full and we prefer to think of it in terms of a prescription – 'Add together all the D numbers with subscripts from 0 to 49 inclusive.' This is still a bit wordy, and the next step of course is to bring in an abbreviation for this prescription. 'Add together' is usually indicated by Σ, the Greek S. This is used because S is the first letter of 'sum'. Some writers simply use the English letter S; for instance, K. A. Kershaw of Imperial College in his book *Quantitative and Dynamic Ecology*. You should be prepared for this use of S, but Σ is (I believe) still the most commonly used abbreviation. When we see this sign, we know we are being instructed to form a sum. The next thing is to examine what is being summed. In our sums above, we had terms such as D_0, D_1, and D_2. Each of these consisted of a capital D with some number written below the line. For 'some number' we naturally use an algebraic symbol such as n, r, or any other letter. We thus reach the abbreviation ΣD_n, which tells us that we are to form a sum, each term of which will be D with some number as a subscript. One thing is still not clear, how many D symbols, and which ones, are to be included in the sum. If we wanted to indicate the sum $D_0 + D_1 + D_2$ we would write $\sum_{n=0}^{n=2} D_n$. The $n = 0$ under the summation sign indicates that we are to start with $n = 0$, i.e. we are to write D_0 first. The $n = 2$ above the sign indicates that we are to go on writing D symbols until we reach D_2; then we are to stop.

In the same way, the number of robins dying before their fifth birthday would be represented by $\sum_{n=0}^{n=4} D_n$ and the number of people dying before 50 by $\sum_{n=0}^{n=49} D_n$.

We are not compelled to start with $n = 0$. If we wanted to write the number of people dying in their seventies, that is, at ages 70 to 79 inclusive, we would write $n = 70$ below the sign and $n = 79$ above it. This would indicate $D_{70} + D_{71} + \ldots + D_{79}$.

Sometimes you see $n = \infty$ written above a summation sign. This means that you are to go on adding for ever; ∞ is the sign for 'infinity'. Thus $\sum_{n=0}^{n=\infty} D_n$ would indicate the number of people or robins that die at any age whatever. You may object that this sign is unnecessary; nobody lives beyond the age of 200; if we wrote $n = 200$ above the sign we would be leaving an ample margin. It is conceivable that advances in medical science may one day prolong life beyond 200 years. By using $n = \infty$ we cover all such possibilities; the instruction is to keep on adding so long as there is anything to add.

The signs just explained are in frequent use but they are extremely inconvenient for the printer. Authors aware of this difficulty are beginning to use forms of instruction which do not involve a lot of little symbols above and below the line; the instructions when to begin and when to stop may be given after the main expression. Thus $D_0 + D_1 + D_2$ may appear as ΣD_n, $n = 0, 1, 2$. The sum which began at 70 and stopped at 79 might appear as ΣD_n, $70 \leqslant n \leqslant 79$. Here the signs explained earlier in this chapter are used. We are told that the acceptable values of n are 70 or more than 70, but must at the same time be less than, or equal to, 79.

CHAPTER SIX

Resistances, Condensers, Springs, and Baths

THE *Mathematical Gazette* for May 1947 reported a presidential address to the Mathematical Association in which Mr W. F. Bushell traced the history of mathematics teaching in England. Mr Bushell pointed out that the problem of a cistern being filled by several pipes occurs already in an arithmetic textbook written by Bishop Tunstall in 1522. He added drily that 'plumbing had to make a considerable advance before this develops into the two taps and waste pipes of nineteenth-century textbooks'. The problem in fact dates from even earlier than 1522. The great Indian mathematician, Bhaskara, born in 1114, discusses it in his treatise *Lilavati*. Rouse Ball, in his *History of Mathematics,* points out that the problem was a practical matter in an age when time was measured with the help of water clocks.

It is perhaps surprising that this ancient problem uses exactly the same mathematics as certain electrical questions, such as resistances in parallel or condensers in series, that belong to our own age. This illustrates well the nature of mathematics, which is not concerned with any particular type of object, but rather with patterns which may recur in the most varied situations.

A typical bath-filling question runs as follows – if tap A will fill a bath in 3 minutes, and tap B, running alone, will fill it in 6 minutes, how long will it take to fill the bath if taps A and B are running together?

This problem, to most people meeting it for the first time, does constitute a puzzle. It is not obvious what calculations should be made in order to reach the answer. The whole trick, in the posing of the puzzle, is concentrated at one point. As a rule, in specifying the flow of water, we express it in some way such as cubic feet per second or gallons a minute, i.e. as quantity per unit time. There is then no problem at all in finding how much two taps supply. If one supplies 3 gallons a minute and the other 4 gallons a minute, together they supply 7 gallons a minute. But the inventor of the problem was careful to specify it the other way, in effect as minutes per gallon. He tells us the time required to fill the bath. The problem ceases to be a problem the moment we realize we must return to the usual way of specifying flow. Instead of saying that tap A takes 3 minutes to fill the bath, we could say that tap A will fill 20 baths an hour. Similarly, tap B will fill 10 baths an hour. Together, then, they will fill 30 baths an hour, which means that a single bath can be filled in 2 minutes.

We have avoided fractions by the device of considering what happens per hour instead of per minute. Without this device we would argue that in one minute tap A supplies one-third of a bath, tap B one-sixth. As one-third and one-sixth add up to one-half, together the taps supply half a bath a minute, and accordingly take 2 minutes to fill a bath. The numbers 3, 6, and 2, representing separate times and combined times, are related (as the argument just used shows) by the equation $\frac{1}{2} = \frac{1}{3} + \frac{1}{6}$. In general, if tap A takes a minutes to fill the bath, and tap B takes b minutes, together they will require t minutes where $\frac{1}{t} = \frac{1}{a} + \frac{1}{b}$. The numbers, a, b, t all appear in the denominators of the fractions. This upside down effect is due to the reversal of the usual procedure for specifying rates of flow.

It would be possible to imagine a problem for which $t = a + b$ would give the answer. It is conceivable that in some industrial process a bath would have to be filled by water coming through tap A, and that as soon as it was full, tap A would be turned off and the water discharged through tap B to some other vessel. It would then take a minutes to get the water into the bath and b minutes to get it out again and the total time, t minutes, would be correctly given by $t = a + b$.

Each form of measuring flow is therefore appropriate to some situation.

We have a similar choice when we are dealing with springs. We may say it requires a 2 lb. weight to stretch a spring 1 inch, or that 1 lb. stretches the spring $\frac{1}{2}$ inch. – pounds per inch or inches per pound. If we have two springs, one of which requires 2 lb. to stretch it an inch, and the other 4 lb., it is fairly evident that 6 lb. will be required for a 1 inch stretch, if the springs are placed side by side as in Figure 54.

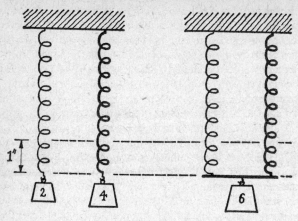

Figure 54

What will be the effect of joining the springs end to end, as in Figure 55? Here we have to use the physical principle that the tension throughout the system is the same. If, for example, the tension in the right-hand spring exceeded that in the left-hand spring, the point where the springs join would find itself pulled to the right. This would lengthen the left spring, increasing its tension, and shorten the right-hand spring, decreasing its tension. The point would continue to move to the right until a position was reached at which the tensions balanced. Accordingly, if we had a pound weight on the end of the combined spring, each part would experience a tension of 1 lb. We want to see what stretches this will produce in the two springs, so for this problem we work

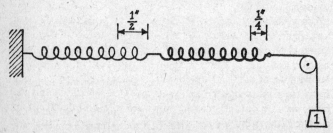

Figure 55

in inches per pound rather than in the system we used in the previous paragraph. The pull of 1 lb. stretches one spring by $\frac{1}{2}$ inch, the other by $\frac{1}{4}$ inch, so the total distance the weight will descend, after being applied to the unstretched system, is $\frac{3}{4}$ inch. To produce a stretch of an inch, the weight would have to be made $1\frac{1}{3}$ lb. Note that the combined spring is now weaker than each of the separate springs; the $1\frac{1}{3}$ lb. to stretch it an inch is less than the 2 lb. and the 4 lb. required for the separate parts.

You will notice that the first question here involved the addition $4 + 2$, the second one involved the addition of the reciprocals $\frac{1}{2} + \frac{1}{4}$. Notice too the mathematical conditions corresponding to the two situations. When the springs are side by side, the distances they stretch must be equal (the tensions in the springs, 2 lb. and 4 lb. weight, are different). When the springs are end to end, their tensions must be equal (the stretches, $\frac{1}{2}$ inch and $\frac{1}{4}$ inch, are different). In solving physical problems, the hardest part is not the actual calculation, but the initial process of analysing the situation and deciding what physical principles apply to it. This is what requires the most practice.

Both questions involved an addition. In the first question, we added tensions which, so to speak, measure the unwillingness of the springs to be stretched. In the second question, we added the stretches which measure the responsiveness of the springs to a given tension. The first arrangement produced a system that was harder to pull out than either spring by itself, the second arrangement a system that was easier.

A spring does not stretch unless a force is applied compelling it to do so. In the same way, a current does not flow in a conductor of electricity unless a voltage is applied to make it do so. We tend to associate voltage with batteries, and this is to some extent useful in visualizing what voltage means. But there can still be a voltage difference between two points, even if neither of them is a terminal of a battery. The comparison with temperature may be helpful. Suppose we have a room in which a furnace is at a temperature of 1,200°. It is cold outside, and a window a long way from the furnace has the temperature 0°. The furnace, and the bitter weather outside, may be compared to the terminals of an electric battery. But the furnace and the window are not the only places where temperature can be measured. If you walked from the window towards the furnace you would experience a steady increase of temperature. It is the same when the terminals of a 100-volt battery are connected by a long piece of wire. If we regard one terminal as being at voltage 0 and the other at voltage 100, then, as a point moved along the wire from the first terminal to the second, its voltage – or potential – would gradually rise from 0 to 100. (The term *potential* is perhaps more usual than *voltage* in this context.)

Many people, if asked to explain Ohm's Law, would probably say something like, 'If a battery of voltage V sends a current of A ampères through a resistance of R ohms, then A = V/R, or V = AR.' The point I want to emphasize is that Ohm's Law applies not only to a complete circuit but also to every piece of it. If somewhere in a complicated piece of apparatus, a current of *A* ampères is flowing through a piece of wire of resistance R ohms, then one end of that wire must have a potential AR volts higher than the other end.

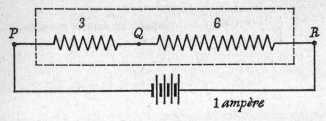

Figure 56

A simple application of this principle is to calculate the effect of resistance in series. In Figure 56 we see resistances of 3 and 6 ohms, through which a battery drives a current of 1 ampère. Accordingly the potential at Q must be 3 volts higher than that at P, since this potential difference drives a current of 1 ampère through the resistance of 3 ohms. In the same way, to make 1 ampère flow through the 6 ohm resistance, the potential of R must be 6 volts higher than that of Q. Combining these results, it follows that the potential of R must be 9 volts higher than that of P. Now suppose the resistances were enclosed in a box (shown by the dotted lines) and that we could only see the terminals P and R.

Finding that we had to put 9 volts across these terminals to make a current of 1 ampère flow, we would be led to believe the box contained a resistance of 9 ohms.

In Figure 57 the dotted lines also represent a box with resistances of 3 and 6 ohms inside, but this time they are connected in parallel. We put 60 volts across the terminals S and T. Now S and T are points at the ends of the 3 ohm resistance, and they are also points at the end of the 6 ohm resistance. The equation $V = AR$ applies to each resistance. To give 60 volts between its ends, the 3 ohm resistance must be carrying a current of 20 ampères.

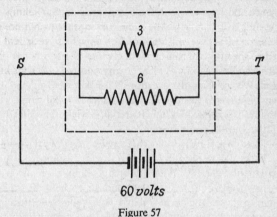

60 volts

Figure 57

Similarly, the 6 ohm resistance must be carrying 10 ampères. The total current delivered by the battery is thus 30 ampères. Now a battery of 60 volts delivers a current of 30 ampères when connected to a resistance of 2 ohms. If we were not able to open the box, our first guess would probably be that it contained a resistance of 2 ohms. In this sense, the resistances in parallel are said to be equivalent to a resistance of 2 ohms.

If you compare the calculations just made with the discussion of the bath problem at the beginning of this chapter, you will find that the figures appearing are identical.

It is not surprising that resistances of 3 and 6 ohms connected in this way are equivalent to 2 ohms, with the number 2 smaller than both 3 and 6. It is rather as though a crowd emerging from a football match or a theatre find two narrow gates side by side. Some will pass through one, some through the other, and they will thus get out more easily than if either gate had been there by itself. The situation in Figure 56, where the resistances are in series, corresponds to the crowd having to pass first through one gate and then through another.

We brought 60 volts in for the same reason that we brought 60 minutes into the bath question – to avoid fractions. If we did not bother to use this device we would naturally consider 1 volt applied across S and T. Then 1/3 ampère would flow in the 3 ohm resistance and 1/6 ampère in the 6 ohm resistance. The total current would be 1/2 ampère, the same current that would flow if 1 volt were applied to a resistance of 2 ohms, which we accordingly regard as the effective resistance of the box.

This argument can be applied to any number of resistances in parallel. We ask the same questions – if 1 volt were applied, what current would flow through each resistance? – what would be the total current supplied by the battery? – what single resistance would lead to the same current?

For example, suppose that three resistances of r_1, r_2, and r_3 ohms in parallel were equivalent to a single resistance of R ohms. What equation gives R? With 1 volt applied, the currents through the resistances would be $1/r_1$, $1/r_2$ and $1/r_3$. The total would be the sum of these. The current that 1 volt would send through R ohms would be 1/R ampères. If these two currents are to be equal, we must choose R to make $\frac{1}{R} = \frac{1}{r_1} + \frac{1}{r_2} + \frac{1}{r_3}$. For example, if we had resistances of 2, 3, and 4 ohms in parallel, the right-hand side of this equation would be $\frac{1}{2} + \frac{1}{3} + \frac{1}{4}$ which comes to 13/12. Thus 1/R = 13/12 and R = 12/13. Thus the effective resistance of this system would be slightly less than 1 ohm.

The analogy with bathroom taps is very close here. If we spoke of 'resistance' in connexion with taps, a tap with high resistance would surely be one that allowed water to trickle through only very slowly; it would take a long time to fill the bath. Accordingly, we might measure the resistance by the length of time required to fill the bath. This corresponds exactly to what is done in measuring electrical resistance. Suppose a resistance of 2 ohms connected to a particular battery allows a certain amount of electricity to pass in 2 seconds. Then, with the same battery, a resistance of 3 ohms would need 3 seconds to pass the same amount of electricity; a resistance of 6 ohms would require 6 seconds. In 1 second then, these resistances would allow $\frac{1}{2}$, $\frac{1}{3}$, and $\frac{1}{6}$ of that amount to pass. Three resistances in parallel are just like three taps running together. To find their combined effect, we add together the *rates at which they allow electricity to pass,* just as, in the corresponding bath problem, we would add the rates for the taps. The apparent complication in the equation is due to the fact that the rate for any resistance is given not by r but by the reciprocal 1/r, and each term of the equation is thus in the form of a reciprocal.

The rate at which water flows through a tap depends on the water pressure. In the bath problem it is tacitly taken for granted that the pressure does not vary. In the electrical problem we have to be quite explicit about it and say, as we did above, that we are considering a specific, fixed voltage applied, such as 1 volt.

Incidentally, in real life, there may be some resistance inside a battery, which affects the current delivered. All our arguments above have assumed that this resistance, the internal resistance of the battery, is so small that it can be neglected.

CURRENT IN ANY NETWORK

The problems we have considered above, of resistances in series and parallel, are so simple that we hardly needed to use algebra to cope with them. The value of a mathematical approach is seen more clearly when we come to more complicated questions. The situation is that in any network, however complicated, made up of any number of batteries and resistances, the way in which current flows is determined by two simple principles, and the actual calculations require only the solution of a number of simple equations.

One of the principles we have already met – Ohm's Law, applied to each part of the circuit. The other principle is, if anything, simpler. Imagine a pipe through which, say, 10 gallons of water are flowing each second. Imagine that at a certain point, this pipe feeds into two smaller pipes, as in a capital Y, and that in one of the smaller pipes 6 gallons a second flows. What will be the flow in the other pipe? This is surely no problem; everyone will answer 4 gallons a second. Exactly the same computation applies to electric currents. If we have a Y-shaped circuit, and 10 ampères flows into it from the base of the Y, and 6 ampères flows out through one arm of the Y, then 4 ampères must flow out of the other arm. We could state the principle thus; at any junction, the rate at which electricity flows in is balanced by the rate at

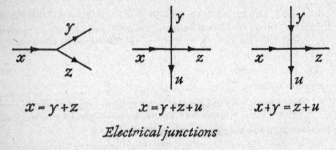

$$x = y + z \qquad\qquad x = y + z + u \qquad\qquad x + y = z + u$$

Electrical junctions

Figure 58

which it flows out. Figure 58 shows various junctions, and the equations that express this principle. All the letters stand for the sizes of currents. It will be seen that writing the equations is very easy.

We will now apply these two principles to the circuit shown in Figure 59. The symbols written inside squares show the potentials of each point in volts; the potentials, 10 and 0 volts, at the top and bottom are given; the other two, indicated by p and q, are unknown. Against each resistance is written a figure, giving its size in ohms. The letters x, y, z, u, w represent the size in ampères of currents, measured in the directions shown by the arrows. Thus we have five unknown currents and two unknown potentials. The first question we must ask is, 'Do we have sufficient information to fix these unknowns?' It can be seen that we do have the number of equations normally required to find seven unknowns, and indeed we can see that this will always be so. For each resistance, we have an equation that expresses Ohm's Law for that resistance. Now for each resistance there is one unknown current pass-

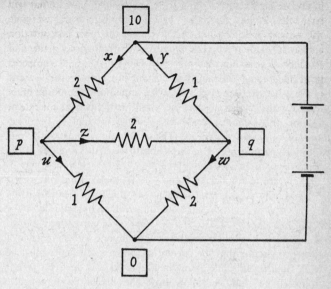

Figure 59

ing through it. So the number of equations we can write down by appealing to Ohm's Law is the same as the number of unknown currents. The unknown potentials occur at junctions, and at each junction we can write an equation expressing the fact that the total current entering that junction equals the total current leaving it. Thus the number of these equations equals the number of unknown potentials. Combining these two considerations, we see that the total number of equations equals the total number of unknowns. As was mentioned earlier, in the section on 'Abundance of Solutions', this does not make it certain that the system of equations will have a solution, nor that there will be just one solution, but at any rate the situation is hopeful and we may go forward. (In fact, in this particular type of problem, as one might guess on physical grounds, everything always works out happily.)

At the top left we see a resistance of 2 ohms through which a current of x amperes flows, the potential dropping from 10 volts to p volts, as we go, in the direction of the current, from one end of the resistance to the other. If we use Ohm's Law in the form $A = V/R$, we find $x = (10 - p)/2$ which means $x = 5 - \frac{1}{2}p$. In the same way we can find the currents in the other resistances in terms of the potentials. The equations we obtain are $y = 10 - q$, $z = \frac{1}{2}p - \frac{1}{2}q$, $u = p$, and $w = \frac{1}{2}q$. The equations at the junctions, expressing 'current in = current out', are $x = u + z$ and $y + z = w$. We can use our earlier equations to substitute in these for x, y, z, u and w. As the result, we get two equations, involving only p and q, and these can be brought to the form $4p - q = 10$, $4q - p = 20$. These equations have only one solution, namely $p = 4$, $q = 6$. Substituting these values in the first five equations, we find $x = 3, y = 4, z = -1, u = 4, w = 3$.

There are a number of comments to make. First, the result $z = -1$ will be noticed. The negative value for z means that the current is flowing in the opposite direction to that shown by the arrow in Figure 59. The actual distribution of current is as shown in Figure 60. It would in fact be clear to anyone with experience of electrical circuits that the current in the wire labelled z must be from right to left. The arrow in Figure 59 was deliberately put from left to right, to show how the algebra automatically corrected this by giving a negative value for z. Here we have an example of the usefulness of negative numbers.

A second point; in Figure 60 it is seen that the battery supplies a current 7 ampères. The number 7 appears in two places on the figure; it shows 7 ampères leaving the battery at the top of the figure and 7 ampères returning to the battery at the bottom. Now of course the current returning to the battery must equal the current leaving the battery. Our analysis made no mention of this fact, and it may well be thought we ought to write an extra equation to embody it. In Figure 59, the current entering the network at the top must be $x + y$ ampères, and the current leaving at the bottom must be $u + w$ ampères. Accordingly the extra equation would be $x + y = u + w$. If we had included this equation, we would have had eight equations for seven unknowns. The situation would be rather like that in the chemistry section, where we had more equations than unknowns. Here, as there, the fact that the equations present a scientifically reasonable problem leads us to expect that the extra equation or equations will be satisfied automatically, and this is indeed what happens.

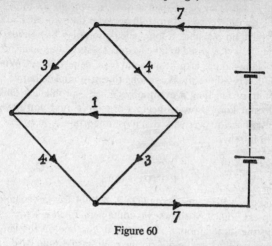

Figure 60

Finally, the arithmetical work in our problem was deliberately made extremely simple, partly to avoid difficulties of printing and partly to avoid distracting the reader by arithmetical complexities. In any real problem, no doubt much more awkward numbers would appear. However it is important to separate difficulties of strategy from difficulties of computation. It is perfectly legitimate, if you are confronted with a problem containing awkward numbers, such as $1 \cdot 576a + 92 \cdot 3b = 117 \cdot 8$, to replace it by a problem containing simple numbers, such as $2a + 3b = 7$. This allows you to concentrate your attention on the strategy to be followed: in which order to use the equations, how to combine them, and so forth. By solving the simplified problem, you give yourself a plan of attack. The identical procedure, with modified numbers, can then be applied to the actual problem. And for this part, the actual computation, you surround yourself with every aid to calculation you can lay your hands on – ready reckoners, slide rules, desk calculating machines, electronic computers; whatever is available. It is a great mistake either to get so tired mentally doing arithmetic that you are unable to plan your grand strategy, or to be demoralized by the appearance of complicated numbers in a question. Computation is a routine matter, something that a machine handles well. A machine takes longer to multiply 2458·71 by 82·35 than it does to multiply 23 by 17, but there is no difference in principle between the two calculations. If you have no machine, a long calculation may be tedious, and you may be well advised to take frequent rests while you are doing it, but being tedious is quite different from being difficult. When a calculation is tedious, it takes a long time to complete but you can be certain of finishing it eventually. When a problem is difficult you do not know how to begin work on it, and you have no certainty that you ever will. It is most important to keep this distinction in mind.

CONDENSERS

An analogy with water containers is helpful for understanding condensers. Figure 61 shows two containers. Each is supposed to contain the same amount of water, but the one on the left has a narrow base, while the one on the right has a broad base. If we made identical small holes in the bases of the two containers, the water would rush out of the left-hand container but only trickle out of the right-hand one. The high column produces a greater water pressure than the low dish. The height of the water is found by dividing the volume of the water by the area of the base.

Figure 61

Condensers are, so to speak, containers for electricity. The *capacity* of a condenser corresponds to the area of the base of a water container. The high column of water corresponds to a high voltage; the shallow water in the dish to a low voltage. The quantity of electricity corresponds to the volume of water. Thus, corresponding to 'height = volume divided by area', we have 'voltage = quantity divided by capacity', or $V = Q/C$.

With condensers, as with resistances, we have the question – what is the effect of placing two in parallel or in series?

Figure 62 shows two condensers, of capacities C_1 and C_2, in parallel. They are placed in a box, indicated by the dotted lines, and two terminals emerge from the box. The question is – is there any single condenser that would produce all the effects we could observe without opening the box?

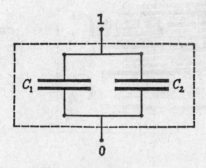

Figure 62

Since the two condensers are connected together, charge can move from one condenser to the other. The first question we must answer is – what equation or physical law decides how the electricity will distribute itself between the two condensers? Now the upper plates are connected together by a wire. By Ohm's Law, a current will flow in this wire if the upper plates are at different potentials. So long as current is flowing, the electricity has not settled down. Accordingly, if and when the electricity reaches its final state, the two top plates must have the same potential. By the same argument, the two lower plates must reach a common potential (not the same as the potential the two upper plates have).

The basic equation for a condenser, $V = Q/C$ can also be written $Q = VC$. If we put $V = 1$, we find $Q = C$, so the capacity of a condenser could be thought of as the quantity of electricity needed to produce 1 volt between its terminals. Now this gives us the solution of our problem. Suppose we arrange for the potential of the lower terminal to be 0 volt and that of the upper terminal 1 volt. We now have $V = 1$ for each condenser. The equation

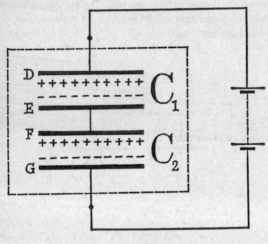

Figure 63

$Q = C$ shows that the charge on the first condenser must be C_1 and on the second condenser C_2. The total quantity of electricity is accordingly $C_1 + C_2$. This is the quantity of electricity that has to be supplied to the terminals of the box to produce 1 volt between them. We would accordingly expect to find a condenser of capacity $C_1 + C_2$ between these terminals, and that is the effective capacity of the two condensers when in parallel.

Figure 63 shows a box containing two condensers in series. In order to get started on this problem, we have to clarify what we mean by a condenser holding a quantity Q of electricity. We do not mean that the condenser as a whole carries an electric charge. Rather we mean that on one plate there is a quantity Q of positive electricity and on the other a quantity Q of negative electricity. Figure 63 shows such charges of positive and negative electricity on each of the condensers. Now we suppose the condensers were uncharged before they were connected to the battery. Where have these charges come from? It is easy enough to see how charges manage to get to plate D at the top and plate G at the bottom, for these plates are connected to the battery, which withdraws electrons from D and supplies them to G. But what about the plates E and F in the middle? These are not connected to anything except each other; there is no way in which electricity can reach them from outside. Accordingly, what must have happened is that the forces of attraction and repulsion exerted by the charges on D and G have caused some electrons to migrate from plate F to plate E, to produce a negative charge on E and an equal and opposite positive charge on F. Thus the quantities of electricity active in the upper and lower condensers must be equal, and here is the physical principle we are looking for.

When the condensers were in parallel, both had to have the same V. Now they are in series, they both have to have the same Q. In the 'parallel' case, we found it convenient to put V = 1, and think of capacity C as measuring quantity per volt. Now we shall find it convenient to do the opposite; take Q = 1, and consider what measures 'volts per quantity'. Putting Q = 1, the equation $V = Q/C$ gives us $V = 1/C$. Thus, just as in the bath problem at the beginning of the chapter, the reciprocal takes the centre of the stage. The reciprocal $1/C$ tells us the voltage produced by unit quantity of electricity. Accordingly, we can find $1/C$ for the box in Figure 63 by asking, 'What voltage between the terminals pulls unit quantity of electrons from the upper terminal and supplies unit quantity to the lower one?' This is readily answered. When there is unit charge on plate D, the voltage of D will be $1/C_1$ above that of E. With unit charge on plate G, the voltage of F will be $1/C_2$ above that of G. E and F are connected together and have the same voltage. Accordingly, the potential of D must exceed that of G by $(1/C_1) + (1/C_2)$ volts. This accordingly must be the reciprocal $1/C$ we set out to find, and so we have the standard formula for condensers in series

$$\frac{1}{C} = \frac{1}{C_1} + \frac{1}{C_2}.$$

CHAPTER SEVEN

Factors and the Remainder Theorem

IN nineteenth century works on algebra you can find tremendously long and difficult expressions that have to be put into factors. Since then there has been a reaction. It is generally agreed that most users of mathematics need not acquire the skill in factoring that was necessary to pass nineteenth century examinations. But it is not at all clear how far the reaction ought to go. Should learners of algebra never hear of factors at all? Should they know one or two simple cases and ignore the rest? Should they practise finding factors at all? It is not easy to answer these questions. Calculations in algebra rarely use just one process. A step involving some knowledge of factors may occur in the middle of an argument, and cause difficulty to someone who is ignorant of factoring. It is extremely hard to say in what branches of knowledge and how often this would happen.

In this state of uncertainty there are two things we can do to help a learner. First, we can set out, as far as we are able, a reasoned case for the learning of factors. This means that we would not simply teach factors because they always have been taught; we would try to collect examples of the situations in which factors arose and the ways in which a knowledge of factors made us more effective. Second, it should be possible to present the main results about factors in a manner that was not too painful and did not take too much time. Then, even if our pupils found they did not use these results too frequently, they would at least have had an interesting experience and not wasted too large a part of their lives.

The expressions $(n + 1)(n + 2)(n + 3)$ and $n^3 + 6n^2 + 11n + 6$ have identical meanings. If you substitute any particular number for n in both of them, the results will always be equal. We pass from the first form to the second by multiplying out, a process which is frequently needed when we are simplifying some expression. We pass from the second to the first by factoring. The question at issue is – why should we desire to do this?

There are various reasons. For one thing, the factor form is often easier to calculate. If you put $n = 8$ in $(n + 1)(n + 2)(n + 3)$ you get $9 \times 10 \times 11$, which is easily found to be 990. Putting $n = 8$ in $n^3 + 6n^2 + 11n + 6$ also gives 990, but takes noticeably longer to work out.

Again, the factor form may reveal some significance which is concealed by the other form. If some calculation leads to the answer $n^3 + 6n^2 + 11n + 6$ this does not at first sight seem particularly striking. But if you notice that this can be put in the form $(n + 1)(n + 2)(n + 3)$, you immediately see that you have a remarkable answer. In fact, it probably indicates that the answer could have been obtained more easily by some other method, which would have led directly to the factor form. It happens to all of us, including the greatest mathematicians, that we do not see the simplest way of attacking a problem. Instead we use some laborious method involving heavy calculations. And then, at the end, unexpectedly everything simplifies, and we are left with a short and simple answer. The form of this answer will sometimes suggest a method by which it could have been found more easily, and we can use the improved method with other problems of the same type. In the chapter 'A Method of Discovery' we shall meet certain results the significance of which will only appear when they are put into factors.

Factors can be useful for checking accuracy. The importance of checking cannot be exaggerated. As we have already mentioned, checking does not interest school pupils much, because nothing depends on the accuracy of their work. Someone who is making calculations for some real and important purpose is continually haunted by the possibility of error. Errors also lead to waste of time; if a slip has been made early in a calculation, all the time devoted to the work has been useless. One therefore develops the habit of looking at the work at every opportunity and asking, 'Is this result reasonable?' As an example of the way in which factors can help us to check a formula we may consider the expression giving the area of a triangle in terms of the lengths of the sides, a, b, c. If A denotes the area of the triangle, we can use either the equation $16A^2 = 2b^2c^2 + 2c^2a^2 + 2a^2b^2 - a^4 - b^4 - c^4$ or the same in factor form

$$16A^2 = (a + b + c)(b + c - a)(a + c - b)(a + b - c).$$

Factors and the Remainder Theorem

Here again, incidentally, the factor form is easier to compute than the other expression, Further, we notice that one of the factors is $a + b - c$. This factor will be 0 if $c = a + b$. Is this reasonable? What does it mean in a triangle if $c = a + b$? Should a triangle for which this holds have area nothing?

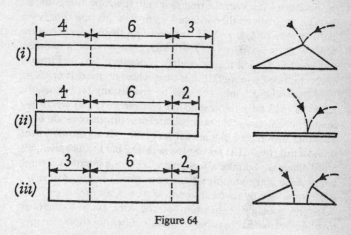

Figure 64

Figure 64 illustrates a way of demonstrating the construction of a triangle. At (i), we see a strip of paper divided into parts 4, 6, and 3 inches long. By bending it at the places indicated, we can bring the ends together and form a triangle, as shown by the sideways view at the right. At (ii) we try to do the same with a strip containing lengths 4, 6, and 2 inches. To make the ends meet, we have to bend the outer parts right over until they lie flat on the middle part. At (iii), with lengths 3, 6, and 2 inches, we cannot make the ends meet at all.

If we take c to be the length of the middle piece and a and b the lengths of the outer pieces, diagram (i) shows a situation in which $a + b$ is larger than c, diagram (ii) shows $a + b = c$, while in (iii) we have $a + b$ less than c.

Accordingly $c = a + b$ indicates the kind of situation seen in diagram (ii) and it is extremely reasonable that we should have area 0 in this case. The presence of the factor $b + c - a$ similarly ensures that A will be 0 when a is the longest side, and is so long that b and c can only reach each other by lying flat against a. In the same way the factor $a + c - b$ looks after the corresponding case when b is the longest side. It is thus reasonable that all three factors should be present in the expression for $16A^2$.

Case (iii) in Figure 64 is an example of a situation where it is impossible to construct the triangle at all. It will be interesting to see what happens to the formula for $16A^2$ in this case. For it we had $a = 3$, $c = 6$, $b = 2$. If we substitute these values we find $16A^2 = 11 \times 5 \times 7 \times (-1) = -385$. Now we cannot find any real number A such that its square is negative. So the formula meets this test quite well. In this case, where no triangle exists to meet the specifications, it refuses to give us any answer at all. Notice incidentally the distinction between case (iii) where we cannot find any value for the area, and case (ii) where we do have a definite answer, but that answer is 0. 'There is no answer' may sound a little like, 'The answer is – *nothing*,' but the meanings are quite different. To take a legal parallel, to the question, 'What crimes have you committed?' the responses 'I refuse to answer' and 'None' are by no means equivalent.

Before leaving the question of the area of a triangle, it may be mentioned that this formula is often given as $A = \sqrt{s(s - a)(s - b)(s - c)}$ where $s = \frac{1}{2}(a + b + c)$. This formula is equivalent to the one we used above, as may be verified by substituting for s, squaring and simplifying.

FACTORS AND ZEROS

In the last section we had a formula for $16A^2$. On geometrical grounds – by considering case (ii) – we saw that $16A^2$ had to be 0 if $c = a + b$. The formula did in fact give 0 because it contained the factor 0.

If we want to make up a formula that will be 0 in certain circumstances, we can achieve this by including a suitable factor. Suppose, for instance, a teacher wants to give a class an equation to solve, but he does not want them to be bothered by difficulties in arithmetic. He thinks it will be good if he can find an equation that has the answers $x = 1$ and $x = 2$. He can find an expression that will give 0 for $x = 1$ by including a factor $x - 1$, for when one factor in a product becomes 0, the whole thing is certain to be

0. To make sure that the expression is 0 for $x = 2$ he can include a factor $x - 2$. He can meet both requirements by choosing $(x - 1)(x - 2)$. When he multiplies this out he gets $x^2 - 3x + 2$, and so he asks the class to solve $x^2 - 3x + 2 = 0$.

I have just said, 'He *can* do this'. But one can make a stronger statement. We suppose he intends to use the simple type of expression usually found in the early stages of an algebra textbook (the technical term is *polynomial*). The position then is not merely that he *can* make the expression 0 for $x = 1$ by choosing one with a factor $x - 1$ but that he *must* do so. It can be proved, and we shall prove it later, that any polynomial which gives 0 when $x = 1$ is substituted must contain the factor $x - 1$. If it becomes 0 when we substitute $x = 2$ it must contain the factor $x - 2$. And quite generally, if it becomes 0 when $x = a$ is substituted, it must contain the factor $x - a$.

For the moment we will not prove this result, but we will assume it to be true and consider some of the consequences. This, incidentally, I believe to be a sound teaching procedure. Young learners of mathematics should know of this result even if they do not know the proof. The proof is not particularly difficult, but a beginner probably would be able to understand this result some time before he reached the stage where he could appreciate the proof. As the result is a very useful one, it is perfectly legitimate and indeed very sensible to tell pupils about it early, and say that they will meet the proof a little later.

The result ties factors and equations tightly together. Saying that an equation has a solution $x = a$ becomes identical with saying that an expression has a factor $x - a$. Some of the things we know about equations can be proved very easily by considering the corresponding statement about factors. Suppose, for instance, we want to find an equation which has among its solutions three different numbers, say 2, 5, and 7. To construct such an equation we must use an expression with the factors $x - 2$, $x - 5$, and $x - 7$. The simplest such expression is $(x - 2)(x - 5)(x - 7)$, which would give us the equation $x^3 - 14x^2 + 59x - 70 = 0$. Our terms of reference were that 2, 5, and 7 were to be among the solutions of the equation: there could be other solutions as well. So we could, if we liked, use the equation $(x - 2)(x - 5)(x - 7)(x - 10) = 0$. Again, if we liked, we could take the equation

$(x - 2)(x - 5)(x - 7)(x^2 + 1) = 0$. I will not multiply out these last two expressions, beyond remarking that the first begins with x^4 and the second with x^5. The point that emerges is this; when we follow the above procedure to construct an equation with three distinct solutions, we are bound to arrive at an equation containing x^3 *or some higher power*. A quadratic equation, $x^2 + px + q = 0$, can never have three separate solutions. The principle is quite general. If we wanted to construct an equation with five different solutions, we should have to put five factors in, and we would be bound to get at least x^5 in the resulting equation. No equation can have five solutions unless it contains x^5 or a higher power. It is convenient to have a short word to specify the highest power occurring in an equation. If the highest power is x^5, we say the equation is *of the fifth degree*. The conclusion we have just reached is that the degree of an equation must always be at least as large as the number of different solutions it has. We can put this the other way round; the number of solutions can never exceed the degree of the equation. A quadratic, of the second degree, can never have three solutions; an equation of the fourth degree can never have five solutions; an equation of the nth degree can never have $n + 1$ different solutions.

A small, but very important and useful note has to be added to this last statement. The need for it can be seen from a particular example. Consider the equation

$$x(x + 1) - 3(x + 1)(x + 2) + 3(x + 2)(x + 3) - (x + 3)(x + 4) = 0$$

The $+$ and $-$ signs separate this into four parts. Each part contains only two factors involving x. Thus, if we multiplied it out we should never get anything higher than x^2. So this is a quadratic equation. But let us start examining what solutions it has. If we put $x = 0$, we get $0 - 6 + 18 - 12 = 0$. If we put $x = 1$, we get $2 - 18 + 36 - 20 = 0$. If we put $x = 2$, we get $6 - 36 + 60 - 30 = 0$. Thus $x = 0$, $x = 1$, and $x = 2$ all seem to be solutions. We have a quadratic with three solutions!

You may like to multiply out and simply the equation, and see what quadratic it is, before reading on. The equation looks a little long, but in fact it does not require much work. One begins, of course, by multiplying out the four separate parts. The only point that requires care is to see that the parts with minus signs are subtracted properly.

*   *   *

When we carry out this work, we find that the first and third parts add up to $4x^2 + 16x + 18$. The second and fourth parts, which have to be subtracted, also add up to $4x^2 + 16x + 18$. Thus, after subtraction, we find 0 on the left-hand side of the equation. We could express this by saying that the equation is $0x^2 + 0x + 0 = 0$. This you can call a quadratic equation if you like; it is of the form $ax^2 + bx + c = 0$, with $a = b = c = 0$. Now of course this equation has an infinity of solutions; whatever number you substitute for x, $0x^2 + 0x + 0$ will give 0. Such equations are often referred to as *identities*. An identity is an algebraic statement that is true for *all* numbers. For example, the statement $(x + 1)^2 = x^2 + 2x + 1$, which we met in Chapter 8 of *Vision*, is an identity.

Accordingly, we have to make clear what is meant by the statement that no quadratic equation can have three solutions. The full statement would be – no quadratic equation $ax^2 + bx + c = 0$ can have three different solutions unless it is the identity $0x^2 + 0x + 0 = 0$, which is true for all values of x. A similar principle applies to equations of higher degree. For instance, the only cubic equation with more than three solutions is $0x^3 + 0x^2 + 0x + 0 = 0$, which has every number for a solution.

This result is very valuable for checking the accuracy of work. In our work above we found $x(x + 1) + 3(x + 2)(x + 3) = 4x^2 + 16x + 18$. Suppose we want to make sure this equation is correct. We can check it by taking $x = 1$. The left-hand side gives 38; so does the right-hand side. This is comforting. It makes it *more likely* that we have not made a slip. But it does not make it certain. The same check applied to $x^2 + 1 = 2x$ would yield 2 on both sides. But $2x$ is certainly not a correct simplification of $x^2 + 1$. The two expressions are in fact equal only for $x = 1$. Our check just happened to light on a thoroughly unrepresentative number. However, it is different if we test the equation at the beginning of this paragraph for three values, such as $x = 0$, $x = 1$, and $x = 2$. Then it ceases to be a question of likelihood and becomes one of certainty. For the equation nowhere involves any power higher than x^2. It represents therefore a quadratic equation. But it has three solutions, which is more than any normal quadratic equation is allowed to have. It must therefore be the exceptional case, an identity true for all values of x. This is how we hoped it would be. There is now no point in making further checks (unless perhaps we feel we may have made a mistake in the arithmetic of the checks themselves). If the checks were correctly done, we have *proof* that the equation is right.

Here a point about language may be made. Some people use the word *proof* as meaning no more than a check. In the English language there is some justification for this; one thinks of phrases such as 'The proof of the pudding is in the eating.' In mathematics however there are two ideas which it is essential to keep clearly separate. A *check* is a procedure which makes it *less likely* that an error has passed undetected. A *proof* is a logical argument which attempts to establish with absolute certainty that some statement is correct. The attempt is not always successful. Arguments which seem entirely convincing in one century are often regarded as unsatisfactory in the next. Even so, the distinction between a check and a proof is an important one to maintain, and it is harmful if the words are used interchangeably.

SOLVING EQUATIONS

It sometimes happens that we meet an equation of which we know one solution. It is then possible to simplify the problem by removing a factor. For example, the equation $x^3 - 18x^2 + 31x - 14 = 0$ is satisfied by $x = 1$. That means the expression on the left side of the equation must have the factor $x - 1$. By long division we find that the other factor is $x^2 - 17x + 14$. Thus the equation may be written $(x - 1)(x^2 - 17x + 14) = 0$. If we take two numbers, neither of which is 0, and multiply them together the result will always be different from 0. So the only way to make the product of two factors 0 is to make one of the factors 0. Accordingly, if replacing x by some number makes our last equation true, it must do it either by causing the first factor, $x - 1$, or the second factor $x^2 - 17x + 14$, to give 0. To make the first factor 0 we must replace x by 1, and we already know that $x = 1$ is a solution, so we get nothing new from this part. To make the second factor 0 we must replace x by a number that makes $x^2 - 17x + 14 = 0$.

That is, we have to solve a quadratic equation. Now there is a definite procedure, not too complicated, for solving quadratic equations. We need not for the moment go into the question of what this procedure is. The only point being made at present is that, by spotting the solution $x = 1$ and separating the corresponding factor $x - 1$, we have succeeded in reducing the problem of solving an equation involving x^3 to the simpler problem of solving an equation that goes no higher than x^2.

You may object that the problem just considered is an extremely artificial one. And indeed it is. I simply made up the equation so that it would have $x = 1$ among its solutions. But it can happen, when we are dealing with a problem arising naturally in reality, that one solution of the problem is obvious while the other solutions are not at all easy to guess. In such a situation the method explained in this section can be very useful. We shall meet such a situation later in this book. (See 'The Doubling Problem' in Chapter 10.)

SOME WELL-KNOWN FACTORS

This section is concerned with certain standard results. The information given here does not, so far as I can see, by itself have any important direct application. Rather it enables us to deal with situations that arise in the middle of some calculation. The first result to be given certainly has a way of cropping up frequently. Some have questioned the value of the others. I would defend their inclusion on the two grounds set out at the beginning of this chapter. They are on occasion useful to know, and, with the treatment here given, very little time is required to deal with them. For the later results arise from exactly the same considerations as the first. A single principle covers the whole business. Seeing that the later results spring from this principle, it seems a pity not to mention them.

DIFFERENCE OF SQUARES

Suppose we want to find the factors of $x^2 - 1$. In view of the close relationship between factors and equations discussed earlier, our problem is essentially the same as that of solving the equation

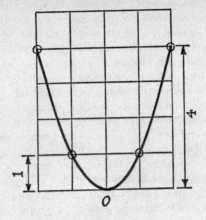

Figure 65

$x^2 - 1 = 0$. By adding 1 to both sides, we obtain $x^2 = 1$. At this stage it may be useful to visualize the graph of x^2. It is as shown in Figure 65. It reaches the height of 1 at the two points with $x = 1$ and $x = -1$. These are the solutions of our equation. The corresponding factors are $x - 1$ and $x - (-1)$. The latter simplifies to $x + 1$. If we multiply these factors together we find

$$(x - 1)(x + 1) = x^2 - 1.$$

In the same way, if we wanted to factor $x^2 - 4$, we could look on the graph for the points with $x^2 = 4$. This would give us the solutions $x = 2$ and $x = -2$, and lead us to the result

$$(x - 2)(x + 2) = x^2 - 4.$$

We could in fact repeat the process with any perfect square replacing 1 or 4. If we introduce the symbol a^2 for the number that replaces 1 or 4, we find ourselves trying to factor $x^2 - a^2$. We look for the points where $x^2 = a^2$. These are $x = a$ and $x = -a$. The corresponding factors are $x - a$ and $x + a$, and the final result is $x^2 - a^2 = (x - a)(x + a)$.

Chapter 8 of *Vision* was entitled 'Investigations'. It gave various exercises in arithmetic which prepared children for, and allowed them to guess, certain results in algebra. We did not include the result $(x - 1)(x + 1) = x^2 - 1$ among these, because at that stage we were avoiding the use of negative numbers and minus signs, and trying to establish only those results which could be easily illustrated by drawing pictures. However it is possible to include in the work of quite young children arithmetical exercises which will prepare them for this algebraic result. We may ask them to work out the following exercise, which closely resembles those on pages 172–4 of *Vision*:

$$2 \times 2 - 1 \times 3 =$$
$$3 \times 3 - 2 \times 4 =$$
$$4 \times 4 - 3 \times 5 = \qquad \text{etc.}$$

In each case they will find the answer 1. How can we illustrate these by pictures? We naturally illustrate 3×3 by a square array with three rows of three objects, and 2×4 as a rectangle with two rows of four objects. How can we make the rectangular array from the square array, and have one object left over? Children

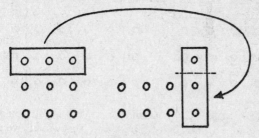

Figure 66

will readily suggest that we slice the top row off the square and stand it up at the side, to form an arrangement that suggests a house with a chimney. (See Figure 66.) Removing the one object that makes the chimney, we are left with two rows of four. Exactly the same procedure shows that 4×4 is just one more than 3×5, and similarly with all the other cases. We have here an informal argument, which I would yet regard as being in essentials a proof, that however far we go with the exercises above the answer will always be 1. As for the algebraic statement of the result, in $3 \times 3 - 2 \times 4$, the number 2 is 1 less than 3, while 4 is

1 more. If we replace 3 by any number x we are led to believe that, starting with x rows of x objects, we can throw 1 away, and still have enough left to make $x - 1$ rows of $x + 1$ objects. This means that $x^2 - 1 = (x - 1)(x + 1)$.

One can similarly devise an exercise such as

$$3 \times 3 - 1 \times 5 =$$
$$4 \times 4 - 2 \times 6 = \qquad \text{etc.}$$

to suggest the result $x^2 - 4 = (x - 2)(x + 2)$. Figure 67 shows that from a square 5 by 5 we can make a rectangle 3 by 7, and still have a 'chimney' 2 by 2 to throw away.

Quite generally, for any whole number a, we can devise exercises to illustrate the factors of $x^2 - a^2$ and a diagram to show the result pictorially.

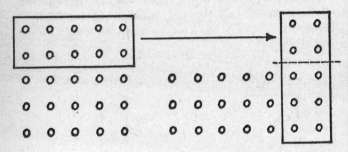

Figure 67

The ideas just discussed can be used for certain types of rapid mental arithmetic, though their main significance does not lie in such applications. For instance, suppose we want to know 18×18. From a square 18 by 18 we can make a rectangle 16 by 20 and have 4 to spare. We can find 16×20 in our heads; it is 320. So 18×18 must be 4 more, 324. We can also work in the opposite direction. For instance, what is 28×32? The rectangle 28 by 32 could be made from a square 30 by 30, again with 4 to spare. So 28×32 must be 4 less than 30×30, that is to say $900 - 4$, which is 896.

We now return for a moment to our first result, $x^2 - 1 = (x - 1)(x + 1)$, to mention an alternative treatment that provides a model for the next piece of our work. It could conceivably happen that someone wished to factor $x^2 - 1$ and realized that this was equivalent to solving the equation $x^2 = 1$ but failed to recognize the solution $x = -1$. This person would see the obvious solution $x = 1$ and accordingly know that $x - 1$ was a factor, This however would be quite sufficient, for he could then divide $x^2 - 1$ by $x - 1$ and find that the other factor was $x + 1$.

Now suppose we wish to find the factors of $x^3 - 1$. This corresponds to solving $x^3 = 1$. But now we can only see the single solution, $x = 1$. Accordingly we have $x - 1$ as a factor, and to find the other factor we must divide $x^3 - 1$ by $x - 1$. The result is $x^2 + x + 1$. Does this have any factors? If it has a factor of the form $x - a$, it will be 0 for $x = a$. So we try to see whether it is ever 0. A natural way to do this would be to draw the graph and see if it ever crosses the horizontal axis. If $y = x^2 + x + 1$, we find the following table:

x	-3	-2	-1	0	1	2
y	7	3	1	1	3	7

In Chapters 11 and 12 of *Vision* we saw that many graphs could be obtained by drawing the graph of $y = x^2$ on tracing paper and then shifting it suitably. The present graph is one that can be so obtained. If you look at the values of y, which are 7, 3, 1, 1, 3, 7, you will notice that these read backwards and forwards the same. The graph is symmetrical and the lowest point occurs midway between $x = -1$ and $x = 0$, since both of these make $y = 1$. Accordingly, we expect the smallest y to occur for $x = -\frac{1}{2}$. Substituting $-\frac{1}{2}$ for x, we find $y = \frac{3}{4}$. Since it seems that y never gets less than $\frac{3}{4}$, it follows that y can never be 0 and so there cannot be any factor of the type $x - a$ in the expression $x^2 + x + 1$. (For the reader familiar with complex numbers, it should be mentioned that this statement should be read as saying that no factor $x - a$ exists with a a real number.)

We thus find that $x^3 - 1 = (x - 1)(x^2 + x + 1)$.

There is nothing special about x^3. If we consider $x^4 - 1$ or $x^5 - 1$ or $x^{17} - 1$ it is evident that each of these gives 0 if we put $x = 1$. So each of them has the factor $x - 1$. We can find the other factor by dividing. The long division would be long indeed for $x^{17} - 1$. However, after working out one or two of the earlier cases, we begin to suspect that we need not actually carry out the division. For a very simple pattern emerges. When we divide $x^4 - 1$ by $x - 1$ we get $x^3 + x^2 + x + 1$. When we divide $x^5 - 1$ by $x - 1$ we get $x^4 + x^3 + x^2 + x + 1$. These expressions may become wearisome to write, but they are simple to specify. We never see anything like $3x^2$ or $10x^3$, all the coefficients are 1. In the last expression we see all the powers from x^4 down to x, in order and added together. Then $+1$ occurs at the end. It might be boring to write down what we would expect to come when $x^9 - 1$ is divided by $x - 1$. We could suggest what it is by writing $x^8 + x^7 + \ldots + x^2 + x + 1$. The highest power here is x^8, the power just below the x^9 that is seen in $x^9 - 1$. We are led to believe that if we divided $x^n - 1$ by $x - 1$, where n could be any whole number, we would get for the other factor an expression containing all the powers from x^{n-1} down to x, and then $+1$ at the end. To prove this result we would have to satisfy ourselves that, if we multiplied that expression by $x - 1$, we would in fact get $x^n - 1$.

It is a curious fact that most algebra textbooks give our last result twice. It appears in two different chapters, and usually there is no mention in either of these that it also occurs in the other. The first chapter, of course, is that on factors. The second is that on geometrical progressions. Geometrical progressions are involved in nearly all financial questions involving compound interest – mortgages, annuities, etc. As we saw in the chapter 'Making the Curved Straight', the mathematics of compound interest is also relevant to other, much more fascinating, situations in nature and in life. An example of a geometrical progression would be the sequence of numbers 3, 6, 12, 24, 48, ... Here each number is twice the number before it. We could write these numbers as 3, 3×2, 3×2^2, 3×2^3, 3×2^4, ... To get the general specification of a geometrical progression we replace 3 by any number a and 2 by any number r. (I imagine r is chosen because it is the initial letter of *ratio*.) In this way we obtain the sequence a, ar, ar^2, ar^3, ar^4, ... This sequence begins with the number a and at each step we multiply by r. Incidentally you may notice that r^2 occurs in the third member of the sequence, r^3 in the fourth, r^4 in the fifth, and so on. The index is always one

Factors and the Remainder Theorem

less than the number that tells us the position in the sequence. If the sequence ends when the nth member is reached, that member will thus contain r^{n-1}. Accordingly, a geometrical progression with n members is specified as $a, ar, ar^2, \ldots ar^{n-1}$. It sometimes becomes necessary to add all of these together. The result is usually described as the sum of the progression to n terms, and may be denoted by S_n. Accordingly we have $S_n = a + ar + ar^2 + \ldots + ar^{n-1}$. There is a factor a in each term here, so we may write the equation as $S_n = a(1 + r + r^2 + \ldots r^{n-1})$. Now the expression inside the bracket is essentially the one we met earlier when we found the factors of $x^n - 1$. Here x has been replaced by r, and also the order of the terms has been reversed. We find 1 at the beginning instead of the end, and the $(n-1)$th power at the end instead of the beginning. But this is unimportant, since the sum of several numbers does not depend on the order in which they are written. Accordingly, we see that the bracket represents the result of dividing $r^n - 1$ by $r - 1$, and so we have $S_n = a(r^n - 1)/(r - 1)$. This is a handy result, particularly if the number of terms in the geometrical progression is large. For instance, if there were twenty terms, and a and r were given particular numerical values, to find the sum of the series in its original form, we would have to work out a, ar, $ar^2 \ldots ar^{19}$ and then add these twenty numbers together. This would be tedious. With the formula we have found, we put $n = 20$ and the only power of r we have to work out is r^{20}. We must then find $r^{20} - 1$ and $r - 1$, which is easy, and divide one by the other. Multiplying the result by a gives us S_n. The work is very much shorter.

It is not necessary that a geometrical progression should have each term larger than the one before. In the chapter on 'Making the Curved Straight' we met several examples of exponential decrease. Suppose a person has for twenty days unwittingly been drinking milk from a farm where radioactive pollution has occurred on a particular occasion. As time passes the radioactivity of the milk will decrease. It might conceivably happen that the amount of radiation affecting this person each day would be nine-tenths of what it had been the day before. If we wanted to find the total radiation absorbed, we would have to sum a geometric progression with $r = 0.9$. In fact, we would have to calculate $(0.9)^{20} - 1$ divided by $0.9 - 1$. Now $(0.9)^{20}$ is about 0.122, and $0.122 - 1$ is -0.878. Accordingly we would have to divide -0.878 by -0.1, which gives 8.78. Thus the total radiation absorbed in twenty days would be about $8\frac{3}{4}$ times that absorbed on the first day. This answer is perfectly correct, but some users of the formula might be thrown off by the stage where we have to divide a negative number by a negative number. Now $(-0.878)/(-0.1)$ is exactly the same as $(0.878)/(0.1)$, so instead of leaving it to each individual user of the formula to make the change of sign above and below in the fraction, most textbooks do this once and for all in the formula. That is, they observe that

$$\frac{a(r^n - 1)}{r - 1} = \frac{a(1 - r^n)}{1 - r}$$

and recommend that the latter form be used when r is less than 1. It should be understood that the first form is *not wrong* even in this situation. It is correct, but it leads to a calculation in which an inexperienced user of the formula may make a mistake.

ANOTHER ASPECT

It is really not at all surprising that $x^4 - 1$ should be $x - 1$ multiplied by $x^3 + x^2 + x + 1$. Consider what happens if we substitute $x = 10$. Then $x^4 - 1$ gives $10^4 - 1$, which is $10,000 - 1$, that is $9,999$. Now it is obvious that $9,999$ has a factor 9 and in fact that $9,999 = 9 \times 1,111$. Now, why does 9 occur here? It occurs because it is 1 less than 10. This suggests – rather faintly, since no general conclusion can be drawn from one particular example – that $x - 1$ may be an algebraic expression lurking behind the 9. The other factor is 1,111, which stands for 1 thousand, 1 hundred, 1 ten and 1, or $10^3 + 10^2 + 10 + 1$, and this in the same way suggests $x^3 + x^2 + x + 1$. If this were all the evidence we had, we could only regard the suggestion as a wild guess. But we do have more evidence. As we saw in Chapter 2 of *Vision*, we work in tens, hundreds, and thousands because we have ten fingers. If we had had some other number of fingers, we could have managed equally well in a system based on that number. In the three-finger system we would find $10,000_3 - 1 = 2,222_3 = 2 \times 1,111_3$. In this system also, the second factor appears as 1,111, and the first factor, 2, is one less than 3, the base of the system. And so it goes on. Whether we use 4, 5, 6, 7, or any other number as the base, we always find the second factor of the form 1,111 and the first factor just one less than the base. That is, if we try a few different bases, we find in each case a result that agrees with the equation $x^4 - 1 = (x - 1)(x^3 + x^2 + x + 1)$. On pages 259-61 of *Vision*, a series of exercises were given designed to suggest such factorings. But we are now in a position to go beyond guesswork to proof. Suppose you had never heard anything about $x^4 - 1$, and that the equation $x^4 - 1 = (x - 1)(x^3 + x^2 + x + 1)$ were put before you. You examine this equation, and you can see that if the right-hand side is multiplied out it is certainly not going to produce any power higher than x^4. That means we are dealing with an equation of the fourth degree. There are two possibilities. Usually an equation of the fourth degree has four solutions. But, as we saw earlier in this chapter, if it has more than four solutions, it must be an identity; it must be true for all values of x. But, by experiments with numbers written in bases 2, 3, 4, 5, and 6, say, we convince ourselves that this equation holds for $x = 2$, $x = 3$, $x = 4$, $x = 5$, and $x = 6$. Here we have more than four solutions. Accordingly the equation must be an identity; it must hold for all values of x. Thus the factoring of $x^4 - 1$ could be both guessed and proved by an investigation starting with the well-known fact that $9,999 = 9 \times 1,111$. By examining the factors of 99, of 999, and so on, in our usual system, and the corresponding results in numbers written with other bases, we could both discover and prove the factoring of $x^2 - 1$, $x^3 - 1$, and the rest. It is in this sense that the opening remark of this section was intended. We remarked that the factoring of $x^4 - 1$ was not really surprising. We have shown that it is related to, and can be remembered with the help of, certain very familiar results in arithmetic.

THE REMAINDER THEOREM

All of this chapter has been based on the belief that solutions and factors are tightly bound together. It is now time to give the logical justification of that belief.

In arithmetic, if we want to decide whether 261 is a multiple of 7 or not, we divide 261 by 7 and see whether any remainder results. We find that 7 goes 37 times into 261 and leaves a remainder of 2. So 261 is not a multiple of 7. On the other hand 7 goes into 91 just 13 times with no remainder, so 91 does have the factor 7. In algebra, too, if we want to find whether one expression is a factor of another, we carry out the division and see whether any remainder results. In certain cases we do not need actually to work through the division. If we are examining some polynomial to see whether it has a factor such as $x - 1$ or $x - 2$, in fact anything of the type $x - a$, there is a very useful rule which tells us straight away what the remainder will be. This rule is known as the Remainder Theorem.

It will help you to understand the Remainder Theorem if you can guess it for yourself. A few exercises are given below which may enable you to do this. They involve division. How to do division in algebra was explained in considerable detail in *Vision*, pages 213–19 and 262–9. The exercises below are in two parts. The first set are simple to do, and suggest the result but perhaps not very strongly or clearly. The second set suggest the result quite clearly but require a little more algebraic skill. They are preceded by a note on a point that arises in the course of the calculation. In both sets of exercises attention should be focused on the remainders that arise.

Exercises

1. Divide x^2 by $x - 1$ and record the remainder.
2. Record the remainder when x^2 is divided by $x - 2$.
3. The same for x^2 divided by $x - 3$.
4. And for x^2 divided by $x - 4$. What do you notice about the remainders in this and the previous questions?
5. In questions 1 to 4 replace x^2 by $10x$, do the divisions and see what you observe about the remainders.

On page 132 of *Vision* we observed that arithmetic may give us a sequence of numbers, but it tends to obscure the rule behind these numbers. If we want to see the rule for the remainder clearly we should not divide in turn by $x - 1$, $x - 2$, $x - 3$, and $x - 4$ but by the general form $x - a$ which covers them all and will lead us directly to the general rule. But of course to do this, we must be able to handle a division of this kind. The point that arises is rather like that on page 109 when we went from the addition $3z + 2z + z = 6z$ to the addition $az + bz + z = (a + b + 1)z$. Here $a + b + 1$ looks so much longer than 6, and yet in any particular case, when we substitute numbers for a and b, $a + b + 1$ does give just a single number like 6. We have to get used to thinking of expressions like $a + b + 1$ as single numbers, doing with them what we would with a particular number like 6, and not worrying about how long they look.

If we are dividing $x^2 + 4x + 5$ by $x - 3$, in the language of page 266 of *Vision* our first step would appear as follows

	Initial amount	$x^2 + 4x + 5$
x rows of $x - 3$ objects use up		$x^2 - 3x$
	and leave	$7x + 5$

The only difficulty here is that of subtracting $-3x$ from $4x$, and this difficulty is one that ought to be overcome before embarking on the kind of work we are doing at present. We may notice that $7x$ contains 7, which is the sum $3 + 4$. If we were commanded not to work anything out, but rather to leave everything in a form which showed how we arrived at it, the last line of the above calculation would have to appear as $(3 + 4)x + 5$. This is the kind of form we have to use if we replace 3 by a, and are dividing $x^2 + 4x + 5$ by $x - a$. For then $3 + 4$ is replaced by $a + 4$, which we certainly cannot work out in the way we work out $3 + 4$ as 7. Our calculation would thus start

	Initial amount	x^2	$+4x$	$+5$
x rows of $x - a$ objects use up		x^2	$-ax$	
	and leave		$(a + 4)x$	$+5$

At our next step we use up $a + 4$ rows of $x - a$, which means $(a + 4)x - (a + 4)a$ or $(a + 4)x - (a^2 + 4a)$. This has to be subtracted from the amount $(a + 4)x + 5$ left after the first step. The remainder is thus $a^2 + 4a + 5$.

Exercises

1. Divide $x^2 + x + 1$ by $x - a$. Record the remainder.
2. Divide $x^2 + 2x + 3$ by $x - a$. Record the remainder.

3. What do you observe in your answers to (1) and (2) and in the remainder found in the text above?

   The answers to all these questions will be discussed after the signs * * * below.

* * *

In the first set of exercises, when we divided x^2 the remainders were 1, 4, 9, 16. These numbers are 1^2, 2^2, 3^2, and 4^2. So, for example, when x^2 was divided by $x - 3$ the result was 3^2. Now 3^2 is what you get if you substitute $x = 3$ in x^2. In the same way, when we divided $10x$ the remainders were 10, 20, 30, and 40, and these are what you get if you substitute 1, 2, 3, and 4 for x in $10x$.

In the second set of exercises, when $x^2 + x + 1$ was divided by $x - a$ the remainder was $a^2 + a + 1$. When $x^2 + 2x + 3$ was divided by $x - a$ the remainder was $a^2 + 2a + 3$. Here we see very explicitly that the remainder is the result of replacing x by a in the expression you are dividing. This last statement constitutes the Remainder Theorem.

The proof of the Remainder Theorem is surprisingly short. In fact to prove it calls for considerably less skill in algebraic computation than is required to guess it by working the second set of exercises above.

The main step in proving this theorem is to get a compact, precise definition of division. In arithmetic we can say that 23 divided by 7 gives 3 with remainder 2. This is a rather rambling, literary type of statement. It would be much easier to build a mathematical argument on it if we could get rid of the words and have simply an equation. Now, as was emphasized on page 262 of *Vision*, a statement about division is essentially a statement about multiplication. Our remark above, about 23 divided by 7, is equivalent to saying that 23 contains three sevens with 2 added, or, as an equation, $23 = 3 \times 7 + 2$. All the information in the original statement is preserved in this equation.

Let us now use a similar equation to explain what we are trying to do when we divide $x^2 + 2x + 3$ by $x - a$. We write the equation

$$x^2 + 2x + 3 = (x - a)(\ldots) + R \qquad \text{(i)}$$

Factors and the Remainder Theorem

The division is an attempt to find what expression should replace (. . .) and what number should replace R to make the above equation an identity, that is, true for every value of x. If we can do this, we shall be in a position to state that dividing $x^2 + 2x + 3$ by $x - a$ gives (. . .) and leaves remainder R.

Now one of our exercises above showed that, in this particular division, the remainder R turned out to be $a^2 + 2a + 3$, which is what you get if you put a for x in $x^2 + 2x + 3$. We know this now, because of the work we did in the exercises, but the result is so simple and striking that it suggests we might somehow get it directly from equation (i) above, without the trouble of doing the division. So we ask ourselves; is there any way of getting from equation (i) a result connecting R with $a^2 + 2a + 3$? Usually when I have asked pupils this question, they have fairly readily found the answer – put $x = a$ in equation (i). After all, there is not much else you could do, if you are going to get from an equation containing $x^2 + 2x + 3$ to one containing $a^2 + 2a + 3$. The moment we try the effect of putting $x = a$, the whole thing falls out. For the equation becomes $a^2 + 2a + 3 = 0 (. . .) + R$. Now it does not matter what (. . .) stands for; when it gets multiplied by 0 the result will be 0. Accordingly we have $a^2 + 2a + 3 = R$, which is what we wanted. This gives us the remainder R, and confirms that it is what we found earlier.

For an informal proof of the Remainder Theorem we now need only observe that what we have done here could be done with any polynomial in place of $x^2 + 2x + 3$. This is all right for an informal discussion, but the writer of a treatise on algebra would not be too happy with our method of explanation. Probably our symbol (. . .) for the quotient would not appeal very much to him. Nor would he be pleased with the idea of carrying through the proof for the particular case $x^2 + 2x + 3$ and then simply saying, 'Of course we could do the same with any other polynomial.' He would like these things set down a little more formally. It may help someone who wants to read mathematical textbooks to see how our discussion would look when formalized and compressed.

We are interested in what happens when any polynomial is divided by $x - a$. So first of all we need an abbreviation for *any polynomial*. A symbol such as P(x) may be used for this. If so, it is important to realize that this is a kind of symbolism entirely different from that used in the greater part of algebra. It does not mean P multiplied by x, nor P to the power x, nor any other arithmetical operation applied to two numbers P and x. It is much more like the abbreviations of everyday life, H.M.I. or I.O.U. P(x) stands for any polynomial involving x. In the same way we could use P(x, y) for any polynomial involving x and y, such as $x^2 + y^2$ for instance. However here we do not need to go beyond P(x). When we divide this polynomial by $x - a$ we get as quotient another expression involving x. For instance, when we divide x^2 by $x - 4$ the quotient is $x + 4$, which involves x. As Q is the initial letter of *Quotient* it seems reasonable to use Q(x) for the quotient. Thus Q(x) replaces the sign (. . .) we had earlier. Accordingly, we replace equation (i) by the equation

$$P(x) = (x - a)Q(x) + R \qquad \text{(ii)}$$

This equation contains in itself the information that when some polynomial, P(x), is divided by $x - a$, we get the quotient Q(x) and as remainder the number R.

When we substitute $x = 4$ in x^2 we get 4^2. Where we saw x before we now see 4. This suggests a symbol to show the result of substituting $x = 4$ in any polynomial P(x), namely P(4). In the same way P(1) would indicate the number obtained by substituting 1 for x and P(2) the result of substituting 2. Thus if P(x) happened to be $x^2 + x + 1$, then P(1) would indicate $1^2 + 1 + 1$, which is 3; P(2) would mean $2^2 + 2 + 1$, that is, 7; P(3) would mean $3^2 + 3 + 1$, that is, 13. If a stands for any number whatever, then P(a) indicates the number you get by substituting $x = a$ in the polynomial P(x).

In this symbolism the proof of the Remainder Theorem is extremely brief. Equation (ii) above completely specifies the situation we are dealing with; it tells us that P(x) is being divided by $x - a$ and what the result of this division is. We now need only say, 'Substitute $x = a$ in equation (ii).' We get

$$P(a) = 0.Q(a) + R = R$$

This tells us what R is, and completes the proof.

A few purely routine exercises may help to fix in mind what the

Factors and the Remainder Theorem

Remainder Theorem says, and what it does for us.

The first exercise is worked to indicate what is expected.

Exercises

1. Find the remainder when x^3 is divided by $x - 10$. *Solution*: by the Remainder Theorem, to obtain the remainder, we merely have to substitute $x = 10$ in x^3. This gives 10^3, that is, $1,000$. Answer: remainder is $1,000$.

2. Find the remainder when x^3 is divided by $x - 2$.

Find the remainders in the following cases:

3. x^2 divided by $x - 5$.

4. $x^2 + x + 1$ divided by $x - 10$.

5. $x^2 + 4x + 11$ divided by $x - 2$.

6. x^{20} divided by $x - 1$.

7. $x^5 - 32$ divided by $x - 2$.

8. $x^3 - a^3$ divided by $x - a$.

SOLUTIONS AND FACTORS

With the help of the Remainder Theorem we can immediately prove the close connexion between factors and solutions. The equation $P(x) = 0$ has the solution $x = a$ if $P(a) = 0$. This is what we mean by a being a solution – that you get a true statement when you replace x in the equation by a. But $P(a)$ also gives the remainder, R, when $P(x)$ is divided by a. When $P(a) = 0$, the remainder is 0 and $x - a$ is a factor. Thus if $x = a$ is a solution, $x - a$ is bound to be a factor.

CHAPTER EIGHT

A Method of Discovery

ONE of the questions often discussed by teachers of mathematics is the place of discovery in the learning of mathematics, as opposed to the system by which the teacher tells the pupil what to do. It seems clear that there must be some sort of balance between discoveries, independently made by the pupils, and facts, told by the teacher. It has required the work of many geniuses over a period of several thousand years to bring mathematics to its present state. Obviously it is asking rather a lot of the average pupil if he is required to think all of this out for himself in the few years available for his schooling. Yet, on the other hand, if lessons consist entirely of telling, the pupil is apt to be left with the impression that mathematics is a mysterious subject which he could never have thought out for himself. The best teaching supplies the learner with information, but allows him to make many of the steps for himself, so he sees mathematics as something that was thought out in the past, and something that he can think out for himself, in some measure, now. There can be no mystery in a result you have discovered for yourself.

The present chapter describes a classroom procedure which seems to possess three assets.

First, it begins with something rather like a conjuring trick, which puzzles the pupils and arouses their interest. Second, it gives an impression of the experience a mathematician has when making a discovery; an initial idea leads to a number of questions and problems, so the work develops and grows in a natural way to a fairly broad theory. Third 'it provides the pupils with a method they can apply to many topics in the syllabus, so that these can be reached by discovery rather than by telling.

It seems best to describe the procedure as it would happen in a classroom. A sign such as * * * indicates that a stage has been reached where the class take pencils and paper and try to solve a problem which has just been posed. Readers may care to put the book down at these places, and see what they can discover before reading on.

THE INITIAL TRICK

A trick suitable to the first lessons in algebra may begin, 'Think of a number'. Our present trick, corresponding to a stage a little further on, begins, 'Think of a quadratic expression'. Thus you might think of $x^2 + x + 1$ or $5x^2 - 3x - 4$ or $2x^2 + 7x + 11$. You must not have x^3 or x^4 or any higher power occurring.

Let us suppose, for example, you choose $x^2 + x + 1$. I then ask you to make the calculation you would make if you were going to draw the graph of $y = x^2 + x + 1$ by plotting the points that come when x is taken in turn to be 0, 1, 2, 3, 4. Accordingly you would make the table:

x	0	1	2	3	4
y	1	3	7	13	21

Now of course, you must not tell me that you thought of $x^2 + x + 1$; if you did, there would be no trick. All you tell me are the five numbers in the y row, namely 1, 3, 7, 13, 21. Now it is always possible that someone makes a slip in arithmetic, so before going further I make a little check, by writing on the board as follows:

1		3		7		13		21
	2		4		6		8	
		2		2		2		

In the top row are the numbers you told me. In the next row are the steps by which these numbers grow; from 1 to 3 is a change of 2, from 3 to 7 is a change of 4, from 7 to 13 a change of 6, and so on. In the same way the third row shows the steps by which the numbers in the second row change. The second row contains 2, 4, 6, 8 with an increase of 2 at each step, so the third row contains nothing but the repeated figure 2. And this is the check. If the numbers in the third row are not all the same, a slip has occurred in the calculation. Incidentally, it often impresses the audience to be told, 'You have made an error' at a time when the formula used still seems unknown.

After making this check, I would immediately say 'The quadratic you chose was $x^2 + x + 1$'.

Occasionally, the trick seems to misfire. The pupil who produced the string of numbers says, 'No, that was not my formula'. Usually what has happened is that the pupil has not followed instructions exactly; instead of finding y for the values 0, 1, 2, 3, 4 of x he has used 1, 2, 3, 4, 5. Needless to say, he must do exactly what is asked if the trick is to work.

The impression given by the instantaneous production of the formula used is that the conjuror has almost miraculous powers of mental arithmetic. In fact, the method used is extremely simple. The teacher now proceeds to do what no conjuror should ever do – repeat the trick again and again, until the class discover how it is done. Plenty of blackboard space is needed, as each member of the class calls out his string of five numbers; the teacher checks them and writes the formula used against them. The class are then asked to examine this evidence, and see if they can guess how the trick was done. There are three things they have to observe, two of which are easy and the third rather harder.

We do not have space here to show all the material that would result. We give three examples only. If you feel inclined, you can supplement this, by choosing a number of quadratics, and working out the numbers they would produce. You can then judge what the class might extract from the evidence before them. The solution appears after the sign * * *.

Numbers given, and check   Formula used

8		11		16		23		32	
	3		5		7		9		$x^2 + 2x + 8$
		2		2		2			

1		−2		−1		4		13	
	−3		1		5		9		$2x^2 - 5x + 1$
		4		4		4			

13		20		33		52		77	
	7		13		19		25		$3x^2 + 4x + 13$
		6		6		6			

* * *

What can be seen in the data above? It is not too hard to observe that, when the formula begins with x^2, the number in the bottom row is 2; when the formula begins with $2x^2$, the number in the bottom row is 4; when the formula begins with $3x^2$, the number in the bottom row is 6. In fact, the bottom row gives twice the number that goes with x^2; in more formal language, it contains twice the coefficient of x^2.

We can also see something by looking at the end of the formula. The first formula ends with $+8$, and we notice that 8 is the first number in the top line. Similarly, the second formula ends with $+1$ and we see 1 at the beginning of the top line. The third formula ends with $+13$ and 13 is also seen at the beginning of the top line. So finding the last number in the formula involves nothing more than copying a figure already on the board. ('The last number' is technically known as 'the constant term'.)

The beginning and the end of the formula are thus given by very simple rules. It is harder to spot how the middle is found. Sometimes very interesting discussions arise in classrooms. Pupils suggest rules, which sometimes work for many of the examples on the board. Then attention is drawn to an example where the rule goes wrong, and a fresh guess has to be made. The procedure gives a good example of scientific method.

To spot the correct rule, we need a little more evidence. Since we are interested in the middle term, it would seem wise to keep the beginning and the end fixed, and vary the middle, and see where effects show in the resulting numbers. For instance, we might take in turn the formulas $x^2 + 2x + 8$, $x^2 + 3x + 8$, $x^2 + 4x + 8$. These would give us the results shown below.

8	11	16	23	32	
	3	5	7	9	$x^2 + 2x + 8$
	2	**2**	**2**		

8	12	18	26	36	
	4	6	8	10	$x^2 + 3x + 8$
	2	**2**	**2**		

8	13	20	29	40	
	5	7	9	11	$x^2 + 4x + 8$
	2	**2**	**2**		

If we look at the numbers on the sloping edge of the table (these numbers are shown in heavy print) we see a regular progression; first 8, 3, 2, then 8, 4, 2, finally 8, 5, 2. The middle number is the one that is changing; it is in turn 3, 4, 5 corresponding to formulas containing $2x$, $3x$, and $4x$. This suggests that the way to get the coefficient of x in the formula is to subtract 1 from the first number in the second row of the table. Now this rule in fact works perfectly well for any formula that begins with x^2, but it goes wrong the moment we try a formula that starts with $2x^2$ or $3x^2$. So we have a part of the truth, and our attention has been focused on the first number in the second row. If we now experiment with formulas beginning $2x^2$, we shall find a slightly different regularity. The number in the middle of the formula (the coefficient of x) now turns out to be always 2 less than the first number of the second row. For instance, in our original data, the formula $2x^2 - 5x + 1$ has -5 in the middle, and this is 2 less than -3, which is found at the beginning of the second row. Similarly, when we go on to formulas containing $3x^2$, we shall find that to get the coefficient of x we have to subtract 3 from the first number of the middle row. In fact, it seems that what we have to subtract is the coefficient of x^2 (which we already know from inspection of the bottom row).

The procedure which the 'conjuror' follows in doing this trick is the following. Suppose he is given the numbers 13, 20, 33, 52, 77. He proceeds to make the check on accuracy and thus obtains the table

13		20		33		52		77
	7		13		19		25	
		6		6		6		

All the information he needs is contained in the three numbers shown in heavy print. The bottom number is 6. He halves this, and gets 3 as the coefficient of x^2. So he writes $3x^2$. Now he has 3 on the board and in his mind. He glances at the 7 in the second row, and subtracts 3 from it. This gives 4, so he writes $+4x$. The board now shows $3x^2 + 4x$. Now he has only to copy the 13 from the top row, to get the complete formula $3x^2 + 4x + 13$.

All you need to do the trick is the ability to divide by 2 and to subtract correctly. However, the procedure is not merely a trick.

A Method of Discovery

It represents a serious method that can be used to solve problems. We will consider three such problems here. Further examples will be found in later chapters.

ARITHMETICAL PROGRESSION

A standard question, found in most algebra textbooks, is to find a sum such as $0 + 1 + 2 + 3 + \ldots + 99 + 100$, that is, the sum of all the whole numbers from 0 to 100, without actually doing the addition. Special methods for dealing with such questions are taught. However, anyone familiar with our trick can discover the answer for himself. The word 'discover' should perhaps be emphasized here; our procedure permits us to arrive at the answer with a fair degree of confidence; it does not however provide formal proof of the correctness of the answer.

Let us examine the first few additions we would make if we were adding the numbers from 0 to 100. We would find in turn

$$0 = 0$$
$$0 + 1 = 1$$
$$0 + 1 + 2 = 3$$
$$0 + 1 + 2 + 3 = 6$$
$$0 + 1 + 2 + 3 + 4 = 10$$
$$0 + 1 + 2 + 3 + 4 + 5 = 15$$

We could make a table like this

x	0	1	2	3	4	5
y	0	1	3	6	10	15

where x denotes the last number in the addition, and y denotes the total. Thus, $x = 3$ means that we are considering the addition $0 + 1 + 2 + 3$, which stops at the number 3, and $y = 6$ gives the result of carrying out the addition.

Is there any simple formula behind the numbers 0, 1, 3, 6, 10, 15 that appear in this table?

* * *

The way we arrived at these numbers does not in the least suggest a quadratic formula, but if we apply to them the procedure used in our trick, we find it works very well. Our check for accuracy would give us

$$\begin{array}{ccccccccccc} 0 & & 1 & & 3 & & 6 & & 10 & & 15 \\ & 1 & & 2 & & 3 & & 4 & & 5 & \\ & & 1 & & 1 & & 1 & & 1 & & \end{array}$$

The numbers in the bottom row are all the same, and this we regard as indicating a quadratic. We apply our rule, without being disturbed by the appearance of fractions. Halving the number in the bottom row gives us $\frac{1}{2}$. Taking this from the first number of the middle row gives $1 - \frac{1}{2}$, which is $\frac{1}{2}$. Copying the first number in the top row gives 0. Thus our three coefficients are $\frac{1}{2}$, $\frac{1}{2}$, 0 and the quadratic formula is $\frac{1}{2}x^2 + \frac{1}{2}x + 0$; there is no point in writing $+0$, so the formula may be given simply as $\frac{1}{2}x^2 + \frac{1}{2}x$. Our original question asked us to go on adding until we reached 100. To answer it, we put $x = 100$ and reach the answer $\frac{1}{2}(100)^2 + \frac{1}{2}(100)$, which works out as 5,050.

The sum of all the whole numbers from 0 to x must certainly be a whole number. It is interesting that the answer, $\frac{1}{2}x^2 + \frac{1}{2}x$, contains fractions. Whatever value you may substitute for x here, you will find you always get a whole number. The proof of this remark can be found by considering what happens if x is even, and what happens if x is odd.

What we have just done is a kind of organized guesswork. We have observed that the formula $\frac{1}{2}x^2 + \frac{1}{2}x$ gives the correct result when x is 0, 1, 2, 3, 4, or 5 (for this is as far as our table went). We have *guessed* that it will continue to give the correct answer when x is larger than 5. A mathematician usually guesses a result first and then tries to prove it. So we might now begin to think whether we can find some argument that will show this really is the right formula.

The answer gives us a very faint clue, for it contains the fraction $\frac{1}{2}$, which suggests that twice the sum might be easier to find than the sum itself.

Now we can picture $0 + 1 + 2 + 3 + 4$, as shown in Figure 68, where we see five columns, containing 0, 1, 2, 3, and 4 squares respectively. The total number of squares is thus $0 + 1 + 2 + 3 + 4$. The shape roughly resembles a triangle. You may recall that the usual way to find the area of a triangle is to take two triangles and

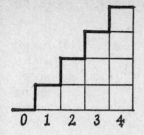 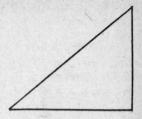

Figure 68

fit them together to make a rectangle. We can do the same with the figure illustrating $0 + 1 + 2 + 3 + 4$. If we take another identical figure, turn it upside down, and put it on top of the first figure, we get the rectangle shown at the left of Figure 69. (The rectangle at the right is not essential to the argument; it simply shows the analogy with finding the area of a triangle.) We have thus built a rectangle with five columns each containing four squares, making twenty in all. Half of this, 10, gives the sum $0 + 1 + 2 + 3 + 4$. We could apply the same idea to finding $0 + 1 + 2 + \ldots + 100$. We would then get a rectangle with 101 columns, each containing 100 squares, or 10,100 squares in all. Half of this is 5,050, the answer we found earlier. The same argument may be applied to find $0 + 1 + 2 + \ldots + x$ where x may be any whole number. You may like to check that this method leads in fact to the answer we had guessed, $\frac{1}{2}x^2 + \frac{1}{2}x$.

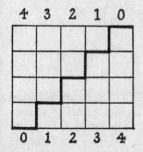 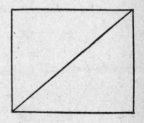

Figure 69

DIAGONALS

Figure 70 shows figures with 3, 4, 5, 6, and 7 sides. The dotted lines are diagonals. The triangle has none of these; the square has 2. The figures with 5, 6 and 7 sides have 5, 9, and 14 diagonals. We thus have a table

x	3	4	5	6	7
y	0	2	5	9	14

where x indicates the number of sides and y the number of diagonals. Is there any simple rule behind these numbers?

*   *   *

If we follow the procedure of the 'trick', we first write down the following figures:

$$0 \quad 2 \quad 5 \quad 9 \quad 14$$
$$2 \quad 3 \quad 4 \quad 5$$
$$1 \quad 1 \quad 1$$

The bottom row contains only the figure 1 repeated, so our data seem to pass the test for a quadratic. People sometimes proceed to apply our usual rule and arrive at the expression $\frac{1}{2}x^2 + \frac{3}{2}x$, but this does not fit. For instance, if we put $x = 4$, this formula would give 14, when it ought to give 2. If we put $x = 3$, the formula gives

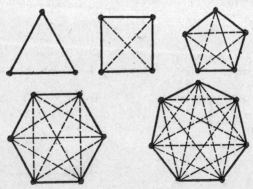

Figure 70
174

9, when it ought to give 0. However the formula fails in an interesting way. You may have noticed that the numbers 9 and 14 which we just found for $x = 3$ and $x = 4$ do occur in the table, but for $x = 6$ and $x = 7$. The formula we have found is out of step by just 3 units. So we could save it, if we modified it by saying, 'First subtract 3 from the number of sides, and put the resulting number for x in $\frac{1}{2}x^2 + \frac{3}{2}x$.' We could even express this by a single formula. If a figure has n sides, subtracting 3 would give $n - 3$, which we would then have to substitute for x. The result would be $\frac{1}{2}(n - 3)^2 + \frac{3}{2}(n - 3)$. A routine calculation (see *Vision*, Chapters 9 and 10) shows that this can be written more simply as $\frac{1}{2}n^2 - \frac{3}{2}n$.

Why did the usual rule give the wrong answer and have to be rescued by a special device? You may remember that, when explaining the trick, we emphasized that instructions must be followed exactly if it was to work, and one of the instructions was that the table must begin with $x = 0$. Now the first table in this section did not do that. It began with the triangle, for which $x = 3$. That is why the resulting formula was just 3 units out of step.

There is another way of dealing with the difficulty. Our data, together with the rows of differences, gave us the following:

x	3	4	5	6	7
y	0	2	5	9	14
		2	3	4	5
		1	1	1	

By examining the pattern of this table, we can extend it to the left, so as to get values of y for $x = 2$, $x = 1$, and $x = 0$. It is easy to extend the bottom line; we merely write more ones. The line above, read from right to left, reads 5, 4, 3, 2, It is reasonable to suppose that it should continue 1, 0, -1. After writing these in, we can write the values of y in such a way that we get a consistent table, i.e. so that each row shows the changes in the row above. In this way we reach the rather surprising table:

x	0	1	2	3	4	5	6	7
y	0	-1	-1	0	2	5	9	14
	-1	0	1	2	3	4	5	
	1	1	1	1	1	1		

If we apply our usual rule to this, we find the formula $y = \frac{1}{2}x^2 - \frac{3}{2}x$, which agrees with what we found earlier.

The negative numbers for y seem rather surprising. Why should a 2-sided figure be thought to have -1 diagonal? Why does a 1-sided figure do the same? Now of course, we cannot draw a region with less than three corners and straight boundaries. The question is really, can we see why our formula should give -1 when we apply it to these cases which geometrically are not very meaningful? To answer this question, we must consider what we mean by a diagonal. It is a line joining two corners of the figure, but it must not be a side of the figure. A square has four corners, and there are six lines joining these. However four of these are required for sides, so the number of diagonals is $6 - 4 = 2$. A triangle has three corners, and there are three lines joining these, but all of these are needed to supply the three sides, so the number of diagonals of a triangle is $3 - 3 = 0$. A figure with four corners has four sides and one with three corners has three sides; on the analogy of this, a figure with two corners should have two sides. But if we put two points on a piece of paper, there is only one line joining them. So we have $1 - 2$ lines available for diagonals. We are in the red. In the same way, a figure with one point should have one side. But there is no line when only one point is there. So the calculation would give $0 - 1$ as the number of lines available for diagonals. Finally, if we have 0 point, we expect 0 side, and the calculation for diagonals would give $0 = 0 - 0$. At this stage we come out of the red.

Incidentally, this discussion shows another way in which we could find the number of diagonals of a figure with x corners. We could start by asking, how many lines are there joining x points? After investigating this experimentally we could subtract from the answer x, representing the number of joins needed for sides.

I am not sure that it is of great value in life to know how many diagonals an n-sided figure has. Rather the discussion is intended as an exercise. Anyone can draw the pictures and count the number of diagonals. It is the method rather than the result that is valuable. Much the same remarks apply to the puzzle which will be described next.

A Method of Discovery

A PUZZLE

You are given a piece of notepaper, and allowed to rule x straight lines on it, in such a way as to divide it into the *largest possible* number of regions. What will this number be?

Most people, presented with this puzzle, would have very little idea how to begin. Our method allows us to arrive at the answer (as a shrewd guess, not as a formal proof) quite quickly. It is important, however, to collect sufficient data. Figure 71 shows

0 line
1 region

1 line
2 regions

2 lines
4 regions

Figure 71

what happens with $x = 0$, $x = 1$, and $x = 2$, that is, when you draw no line, one line, or two lines. You get one region, then two regions, then four regions. A very natural guess is, '1, 2, 4; the number of regions doubles each time you draw a line'. But as soon as you draw the next line, you see this will not work. Whatever you do with three lines, you cannot divide the paper into more than seven regions (Figure 72.)

Guessing is a most important part both of mathematics and of science. You should learn to guess and practise guessing, but

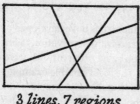
3 lines, 7 regions

Figure 72

guesses should be based on sufficient material and we should never forget that they are guesses. It is always possible that some unexpected fact will turn up one day and show that the guess is incorrect, or perhaps that it needs to be modified in some detail.

I will leave to you the business of drawing figures with four lines and five lines, and guessing what the correct formula is.

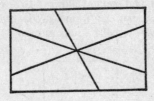

Figure 73

The exact terms of the puzzle should be noted. We are asked to find the *largest* number of regions into which x lines can separate the paper. It sometimes happens that someone draws, say, three lines as in Figure 73 and obtains only six regions. In this figure, three lines pass through a point, which is not the most effective arrangement. If one of the lines were shifted a little, a small triangle would appear in the middle, and we should have an extra region. Clearly we shall get a very irregular table if for some values of x the entries show the maximum number of regions, while for other values we have not made the best use of our opportunities. Such a table will not lead us to the solution we are looking for.

WILL IT ALWAYS WORK?

Many pieces of mathematics have developed along the lines of the topic of this chapter; certain observations were made; methods of calculation, based on these observations, were devised and found to work quite well. The most notable example was perhaps $\sqrt{-1}$, introduced into mathematics around 1575 and used until about 1800 without anybody knowing exactly what it was or why it worked. Naturally, efforts were made to explain it and to prove that its use was justified. So too with our present subject. Can we be certain that our trick will always work? Might there lurk somewhere an exceptional quadratic formula for which it failed? Such considerations lead us to look for a logical proof of the soundness of our method.

Right at the beginning of algebra we considered the trick; think of a number, add 3, double, subtract 4, divide by 2, take away the number you first thought of. This trick evaporates if some sophisticated member of the audience says, 'I think of any number x. The results I then get are $x + 3$; $2x + 6$; $2x + 2$; $x + 1$; 1. I see, the answer must always be 1.' Since x stands for any number whatever, we know at once that the trick works for every number there is.

Our aim is now to use algebra in such a way that we cover every possible quadratic at one blow. Now a quadratic depends on three numbers. Someone, for instance, who chooses the quadratic $5x^2 + 17x + 8$ has decided to put 5 with x^2, 17 with x, and 8 at the end. He might have chosen any three numbers instead of these. We can cover all possibilities if we let a stand for the first number, b for the second, and c for the third. The formula chosen is then $ax^2 + bx + c$.

A mistake beginners sometimes make is to say, 'He can choose any number to go with x^2, any number to go with x, and any number at the end. So let a stand for *any number*. The formula chosen can be represented by $ax^2 + ax + a$.' Why is this wrong? Or perhaps it would be fairer to ask – why will this not fit in with the system of algebraic shorthand that everyone uses today? At various places in *Vision*, we used the idea that a might stand for a number chosen by Alf, b for one chosen by Betty, and c for one chosen by Charles. Now $ax^2 + bx + c$ shows the three numbers, a, b, c, chosen independently by the three children. But $ax^2 + ax + a$ gives a monopoly to Alf. He picks a and this number is written in each of the three places. If Alf picks 2, the formula will be $2x^2 + 2x + 2$. If he picks 11, it will be $11x^2 + 11x + 11$. With $ax^2 + ax + a$, there is no way of bringing any variety into the numbers. It is impossible to get a formula such as $5x^2 + 17x + 8$.

Accordingly we choose $ax^2 + bx + c$ as something that will cover all possible choices. The next thing to do is to see what would be written by someone who chose this expression and then followed the instructions of the trick. This has to be done fairly slowly and carefully, and the work calls for a fairly large sheet of paper or blackboard. The first question is: what would be the five numbers this person would report to the conjuror? You may like to answer this before reading on.

*   *   *

The trick begins; put $y = ax^2 + bx + c$ and make a table showing the values of y for $x = 0, 1, 2, 3, 4$. Learners often seem to make errors in doing this, so it should be done with care. For example, if we are replacing x by 3, then x^2 must be replaced by 9. Accordingly $ax^2 + bx + c$ would become $a9 + b3 + c$, which is usually written $9a + 3b + c$. (The step is justified, since a times 9 is the same as 9 times a). We do similar work with the other values. When $x = 0$, we get $0a + 0b + c$; there is no point in writing 0 times a, which is simply 0, so $x = 0$ just gives c. Accordingly we reach the table

x	0	1	2	3	4
y	c	$a+b+c$	$4a+2b+c$	$9a+3b+c$	$16a+4b+c$

The five entries in the bottom row would be reported to the conjuror. What is the conjuror's next step? Can you carry it out?

*   *   *

The conjuror checks for accuracy, by forming rows of differences, that is, by subtracting each number from the one that follows it. Applying this to the five reported numbers, we get this table:

$$c \quad a+b+c \quad 4a+2b+c \quad 9a+3b+c \quad 16a+4b+c$$
$$a+b \quad\quad 3a+b \quad\quad 5a+b \quad\quad 7a+b$$
$$2a \quad\quad 2a \quad\quad 2a$$

Accuracy does check. With a quadratic, we should have the same number all the way along the bottom row, and we do; everywhere we see $2a$.

If we apply the method of the trick, will it lead us back to the formula $ax^2 + bx + c$ that was actually chosen?

*   *   *

Yes. The method directs us to take half the number in the bottom row, and put it with x^2. As half the number in the bottom row is a, this gives us the correct beginning, ax^2. Our next step is to subtract the number so found from the first number in the middle row, that is, to subtract a from $a + b$. This gives b, which is correct for the number that goes with x. For the constant term, we simply have to copy the first number in the top row. This is c, which again is correct. Accordingly, the method gives $ax^2 + bx + c$. which is the right formula. This covers all possible cases, so we need have no fear, whether we are using this method for a trick or for some more serious purpose, that it may one day let us down.

Incidentally, the approach just used would have helped us if we had failed to guess the rule for finding b, the coefficient of x. By looking at the last table above, we could see $a + b$ at the beginning of the first row, so we would know b could be found by subtracting a from this. In fact, if we were sufficiently experienced in algebra, we could have come directly to this table, and seen how to find a, b, and c from it without any guesswork at all. However, for beginners in algebra, it is well worth while to follow the route we have taken in this chapter.

All the information needed to fix the formula came from the three numbers on the sloping edge of the table. To bring this out in terms of formal algebra, we might write the table like this

The stars here represent numbers we are not interested in at the moment. The symbols p, q, r denote the numbers on the sloping edge. These numbers, of course, the conjuror knows. The numbers he does not know, and is trying to find, are a, b, and c.

If we compare the last two tables, we see that the numbers along the edge will agree only if $c = p$, $a + b = q$, and $2a = r$. Here we have three equations for three unknowns, a, b, c. These equations are very easy to solve. The solution $a = \frac{1}{2}r$, $b = q - \frac{1}{2}r$, $c = p$ simply expresses the rule we have been using throughout.

This discussion of a formal, algebraic approach will prove helpful in the next section.

GENERALIZATION

There are still one or two questions that might be asked about our method for identifying quadratic formulas, but in the main this matter has been dealt with. A mathematician lives by solving problems. Once a problem has been solved, he becomes unemployed and has to look for another problem. So our next question is – what shall we do now?

The answer is not far to seek. So far we have been working under a severe restriction; no power above x^2 must appear in the formula. So it is natural to ask, can a similar method be found for formulas that involve x^3? After that has been answered, we shall begin wondering whether it would matter if x^4 and x^5 or even higher powers were present.

Usually, when venturing into new territory, we find it wise to consider the simplest possible problem, So let us see what happens with the formula $y = x^3$. If we put $x = 0, 1, 2, 3, 4, 5$, in $y = x^3$ we obtain the following table for y and the rows of differences,

0		1		8		27		64		125
	1		7		19		37		61	
		6		12		18		24		
			6		6		6			

Here again we find the bottom row contains the same number again and again, but one more row is needed to reach this. With quadratics we had the row for y and then two rows of differences; with $y = x^3$, we have the row for y and then three rows of differences. Most people immediately make the guess that any cubic formula will do the same – that is to say, any formula of the type $y = ax^3 + bx^2 + cx + d$. Nor do they stop there; they go on to guess that a formula that goes up to x^4, and is thus of the type $y = ax^4 + bx^3 + cx^2 + dx + e$, will give four rows of differences, the last of which will contain a single number indefinitely repeated. In fact, they leap to a general conclusion; if the highest power occurring is x^n, then there will be n rows of differences under the row for the values of y.

It is remarkable how the human mind jumps to conclusions.

A Method of Discovery

We have made a fairly thorough study of quadratic formulas. We have studied one very special cubic equation, $y = x^3$. On the strength of this, we offer to say what happens for any polynomial whatever, however high the powers of x that may occur in it. In this case, at any rate, our boldness is justified. What we guess is in fact perfectly correct.

We now have a very simple way of checking whether a polynomial formula fits a set of numbers. Suppose we are told that, for $x = 0, 1, 2, 3, 4, 5$ the corresponding values of y are 11, 17, 33, 71, 143, 261. Is there a simple formula behind these numbers? We set these numbers out and find the rows of differences:

$$
\begin{array}{ccccccccccc}
11 & & 17 & & 33 & & 71 & & 143 & & 261 \\
& 6 & & 16 & & 38 & & 72 & & 118 & \\
& & 10 & & 22 & & 34 & & 46 & & \\
& & & 12 & & 12 & & 12 & & &
\end{array}
$$

All the numbers in the bottom row are the same, so we stop at this point. We have the numbers for y and beneath them we have three rows of differences. This is an indication that we are dealing with a formula that goes only as far as x^3. So we expect something of the type $y = ax^3 + bx^2 + cx + d$ to fit these numbers. But which formula? What particular numbers should replace a, b, c, d? What can you find out about this ?

* * *

There are several ways of answering this question. Your method may be different from mine, but still be perfectly good.

We can begin by making the table that corresponds to the formula $y = ax^3 + bx^2 + cx + d$ and then trying to make it coincide with the table of numbers given above. If we put 0, 1, 2, 3, 4 in turn for x, we find the corresponding values for y are d; $a + b + c + d$; $8a + 4b + 2c + d$; $27a + 9b + 3c + d$; $64a + 16b + 4c + d$. Putting these in the row for y and calculating the rows of differences we obtain the table

$$
\begin{array}{ccccccccc}
d & & a+b+c+d & & 8a+4b+2c+d & & 27a+9b+3c+d & & 64a+16b+4c+d \\
& a+b+c & & 7a+3b+c & & 19a+5b+c & & 37a+7b+c & \\
& & 6a+2b & & 12a+2b & & 18a+2b & & \\
& & & 6a & & & 6a & &
\end{array}
$$

On the analogy of our earlier work with quadratics, we guess that all the information we need will come from the numbers along the sloping edge – those shown in heavy type above. These correspond to the numbers 11, 6, 10, 12 in the table we are trying to fit.

Thus we have the equations

$$d = 11, \qquad a + b + c = 6, \qquad 6a + 2b = 10, \qquad 6a = 12.$$

These are easily solved. The last one tells us a must be 2. This means that in the third equation we must have $b = -1$. Then, to make $a + b + c = 6$, we must choose $c = 5$. And the first equation tells us directly $d = 11$. Accordingly, if there was a cubic formula behind the numbers given, it must have been $y = 2x^3 - x^2 + 5x + 11$, and this in fact does fit them.

The same method will work for any cubic formula. Suppose the table has the form

The bottom row contains the number s repeated, which we believe is a sign we are dealing with a cubic. The numbers p, q, r, s along the edge are the ones that interest us (they replace 11, 6, 10, 12). The remaining numbers we are not concerned with. They are represented by stars, simply to indicate the rows of the table. In any particular example, we would know what p, q, r, s were.

The four equations for the unknown numbers a, b, c, d would then be

$$d = p, \qquad a + b + c = q, \qquad 6a + 2b = r, \qquad 6a = s.$$

We solve these exactly as we solved the corresponding equations in the particular example we worked just now. The last equation shows $a = \dfrac{s}{6}$. Substituting for a in the third equation, we find $b = \frac{1}{2}(r - s)$. We substitute for both a and b in the second equation, and find after a little arithmetic $c = q - \frac{1}{2}r + \frac{1}{3}s$. The first equation gives $d = p$ without any work at all.

Now that we have expressions for a, b, c, d we can substitute these in the equation $y = ax^3 + bx^2 + cx + d$. We find

$$y = \left(\frac{s}{6}\right)x^3 + \tfrac{1}{2}(r - s)x^2 + (q - \tfrac{1}{2}r + \tfrac{1}{3}s)x + p.$$

You can test the accuracy of our work by applying this result to the particular cases we had earlier. For $y = x^3$ the numbers along the edge were 0, 1, 6, 6. If you put these values for p, q, r, s in the equation above, you should get something which, when simplified, gives $y = x^3$. In our second example, p, q, r, s were 11, 6, 10, 12. If you substitute these values you should get $y = 2x^3 - x^2 + 5x + 11$ for the formula behind the table, since this was the result we found earlier.

In a way, our problem is solved. We now have a rule for finding what cubic equation lies behind a table. It is not anything like as simple as the rule for quadratics, and it does not lend itself to use as a trick. Still, it tells us what we need to know. And yet, in a way, it is disappointing. For our task is not finished. We still have the job of finding the rule for equations containing x^4, and after that for equations with x^5 in them, and so on, for ever. In such an apparently endless investigation our hope is always that at some stage we may notice a pattern which will allow us to guess what happens in all the different cases. So we were not investigating the cubic purely for its own sake. We were hoping that the answer for the cubic would exhibit some feature that would provide a clue to all the later cases, for the further we go in this investigation, the heavier the work becomes. With the cubic we had to solve four equations for the four unknowns a, b, c, d. When x^4 is handled, we shall have five equations for five unknowns; with x^5 there will be six equations for six unknowns, and so it will go on.

Now there are some signs for hope in the answer we have found. The coefficient of x has turned out to be $q - \tfrac{1}{2}r + \tfrac{1}{3}s$, in which there are signs of an orderly progression. We see the numbers 1, $\tfrac{1}{2}$, $\tfrac{1}{3}$ and it seems reasonable to hope that when we deal with the equation containing x^4, we may see the numbers 1, $\tfrac{1}{2}$, $\tfrac{1}{3}$, $\tfrac{1}{4}$ in this place. The signs in $q - \tfrac{1}{2}r + \tfrac{1}{3}s$ are $+ - +$. This is not much to go on but at least it allows us still to hope that we shall find plus and minus signs occurring alternately when we come to the later cases. There is just enough pattern in our answer to make us feel that at this stage it may be worth while to stop and search for a pattern rather than to continue calculating fresh results.

Some pupils carefully examine the individual coefficients and try to guess rules that will give these. A certain amount can be learnt this way, but it is not, I think, the best procedure to follow. Rather it is good to examine the answer for the cubic as a whole. It does in fact contain a very simple and very striking pattern, which is totally hidden by the present form of the answer.

In searching for this pattern, we must be clear what we are trying to do. The answer we have found is in fact the correct answer. We cannot alter it or replace it by some other answer. Our problem is simply to *rearrange* it in a way that will bring out its inner simplicity. You may like to try this for yourself first of all, without any help. If after a certain time you do not feel you are getting anywhere, you may like to read the hint which is given, in the form of a question, below this break in the print.

*   *   *

We are seeking to rearrange the expression for y. At present it is arranged by powers of x, that is, we have grouped together all the terms that contain x^3, all those that contain x^2, and so on. Is there any other principle by which the arrangement could be organized?

*   *   *

Really, once the question has been posed in this form, our choices are pretty limited. Only three ingredients appear in the expression for y — the powers of x, various fractions such as $\frac{1}{2}$, $\frac{1}{3}$, $\frac{1}{4}$, and the symbols p, q, r, s. The equation is already arranged by powers of x, so nothing new is to be found from this possibility. It would be possible to group together all the terms in which the fraction $\frac{1}{2}$ appears. This does not seem a particularly fruitful idea. Finally, it would be possible to group together the terms that contain p, those that contain q, and so on. This would give us

$$y = p + qx + r\left\{\frac{x^2 - x}{2}\right\} + s\left\{\frac{x^3 - 3x^2 + 2x}{6}\right\}$$

You may think this is not much of an improvement, but in fact there is something to notice about the expressions in the brackets. What is it?

*   *   *

A Method of Discovery

Both expressions have simple factors, for $x^2 - x = x(x - 1)$ and $x^3 - 3x^2 + 2x = x(x - 1)(x - 2)$. Thus with p we have simply the coefficient 1; q has the coefficient x; the coefficient of r contains $x(x - 1)$ and that of s contains $x(x - 1)(x - 2)$. Anybody can guess how this sequence continues. When we go on to the formula containing x^4, we shall expect to find $x(x - 1)(x - 2)(x - 3)$ occurring, and indeed we do.

The only thing remaining to be settled concerns the numbers that occur in the denominators of the fractions. The terms p and qx do not involve fractions. If we wanted to drag denominators in somehow, we could write them as $\frac{p}{1}$ and $\frac{qx}{1}$. Thus the denominators in the whole expression for y would be 1, 1, 2, and 6. This is not very much evidence on which to base a guess. If you have sufficient energy, you can work out the formula for y to fit an equation in which the highest power is x^5. On the basis of this work you may be able to guess the rule for the denominators.

* * *

If the formula involving x^5 is worked out, it is found to contain the denominators 1, 1, 2, 6, 24, 120. A rule can be observed here if your attention is drawn to the ratio of each number to the one before it. Thus 120 is 5 times 24; 24 is 4 times 6; 6 is 3 times 2; 2 is 2 times 1. Thus 120 is $5 \times 4 \times 3 \times 2 \times 1$. 120 is known as *factorial* 5. It is found by multiplying together all the whole numbers from 1 to 5 inclusive. The product is usually indicated by dots rather than multiplication signs. In the past factorial 5 was often indicated by the symbol $\lfloor 5$. The line under the 5 was inconvenient for the printer, and in recent books an exclamation mark is always used instead. Thus 5! means 1.2.3.4.5, which is 120.

Thus the pattern of our result for the cubic is fully brought out by writing it in the form

$$y = p + q \left\{\frac{x}{1}\right\} + r \left\{\frac{x(x - 1)}{1.2}\right\} + s \left\{\frac{x(x - 1)(x - 2)}{1.2.3}\right\} \quad \text{(i)}$$

The table for a cubic contains the y row and then three rows of differences. The letters p, q, r, s indicate the numbers at the beginning of these rows. If we wanted to deal with an equation in-

The Search for Pattern

volving x^4, our table would contain the y row and then four rows of differences. We would need p, q, r, s, and t to indicate the initial numbers. To discover the equation behind the table, we would use a formula similar to equation (i) above, but it would contain t as well as p, q, r, s. I imagine most readers will guess the coefficient of t; it is

$$\frac{x(x-1)(x-2)(x-3)}{1.2.3.4}.$$

If we had to deal with equations containing x^5 or x^6 or even higher powers of x, we would use a formula similar to (i), but containing extra terms showing the same pattern.

We thus have an extremely powerful device – a way of identifying the equation behind any sequence of numbers, provided only that the equation involved is a polynomial. It does not matter how high the powers of x that occur in this polynomial.

One restriction should be emphasized. In most scientific experiments, errors of measurement are involved. If a scientist draws a graph, it does not usually go through the points that represent his experimental results, but so close to them that he feels it gives a reasonably good fit. There are mathematical procedures for finding the best curve of a given type to fit data which are known to be inexact. The work we have done does not apply to situations involving errors. We have been concerned with fitting a polynomial *exactly* to certain sets of numbers. Thus the applications of our method are largely to mathematical problems and to practical problems that involve counting rather than measuring. The mathematical problems however are important. The expressions

$$\frac{x}{1} \qquad \frac{x(x-1)}{1.2} \qquad \frac{x(x-1)(x-2)}{1.2.3}$$

of the type that appeared in equation (i) will be met in a variety of mathematical situations. They are known as the Binomial Coefficients. It is good to think about them from time to time, until they become so familiar that they are instantly recognized if they appear part way through some piece of work.

A Method of Discovery

Exercises

1. Find the formulas that give the following sets of numbers corresponding to $x = 0, 1, 2, 3 \ldots$. (Answers should be simplified.)

   (i) 0, 0, 6, 24, 60, 120.   (iii) 0, -1, 4, 27, 80, 175.

   (ii) 0, 1, 4, 10, 20, 35.   (iv) 0, 0, 0, 1, 4, 10.

2. Earlier we found that $y = 0 + 1 + 2 + \ldots + x$, where x must be a whole number, could be written more shortly as $y = \frac{1}{2}x^2 + \frac{1}{2}x$. Find short formulas for the following:

   (i) $y = 0^2 + 1^2 + 2^2 + \ldots + x^2$;

   (ii) $y = 0^3 + 1^3 + 2^3 + \ldots + x^3$.

3. If there were three stations on a railway line, six kinds of single ticket would be printed; A to B, A to C, B to A, B to C, C to A, C to B. Make a table showing how many tickets would be needed if there were 0, 1, 2, 3, 4, 5 stations. (With 0 or 1 station, no tickets would be needed.) From the table, find a formula for the number of tickets needed on a railway with x stations, where x may be any whole number.

4. Make a table giving $y = (x - 1)x(x + 1)$ for $x = 0, 1, 2, 3, 4, 5$. Compare your table with question 1(i). Does this suggest to you that there are two forms in which the answer to 1(i) could be given?

5. Make a table giving $y = x(x + 1)(x + 2)$ for $x = 0, 1, 2, 3, 4, 5$. Compare your table with question 1(ii). Is there any connexion between the two tables? Does anything here suggest an alternative form for the answer to 1(ii)?

6. Tournaments are sometimes organized in which every player plays every other player once. Suppose that if x players enter, there are y games played. Work out y for $x = 0, 1, 2, 3, 4, 5$, and find a formula for y. Does this question have any connexion with any earlier question?

CHAPTER NINE

Discovering the Binomial Theorem

IF you ask someone to write a summary of what is learnt in school geometry, you will probably get a list of theorems: Pythagoras Theorem; the angles at the base of an isosceles triangle are equal; the sizes of the angles in any triangle add up to 180° – results of this kind. But if you ask for a similar summary of school algebra, you usually get a list of processes; how to add, how to subtract, how to multiply, how to divide, and so forth, – very much the same list that you might get for arithmetic, but of course with the understanding that x and y will be involved. There seem to be very few things in algebra that learners think of as theorems. Two results however commonly receive this title – the Remainder Theorem, which we have already met, and the Binomial Theorem, which is the subject of this chapter.

I remember a student who told me he did not really understand the Binomial Theorem. I asked him a few question, which he answered correctly, and I told him he had nothing to worry about. However he was still unhappy; he felt he was missing something. It seemed to me the reason was that textbooks put the Binomial Theorem with capital letters. He perhaps had an association with bibles and prayer books where the names of supernatural beings appear with capital letters. He seemed to feel that the Binomial Theorem ought to be something of profound philosophical importance, not just the little rule which in fact he knew and understood perfectly well. But that is all it is. He was puzzled because he was looking for something that was not there.

Most learners of algebra are familiar with such results as $(x + y)^2 = x^2 + 2xy + y^2$ and $(1 + x)^2 = 1 + 2x + x^2$. The coefficients in both of these are 1, 2, 1. If we go on to cubes, we find $(x + y)^3 = x^3 + 3x^2y + 3xy^2 + y^3$ and $(1 + x)^3 = 1 + 3x + 3x^2 + x^3$. Both of these contain the coefficients 1, 3, 3, 1. The Binomial Theorem is a rule for writing down the numbers that occur in any power of the form $(x + y)^n$ or $(1 + x)^n$. These numbers are also significant in other connexions. Genetics is one field where they occur. For instance, if you cross white and red shorthorn cattle, the offspring are roan. If you then mate roan cows with roan bulls, you can expect to get pure white, roan, and pure red offspring in the ratio $1:2:1$ – a very well known illustration of Mendelism, and using the same numbers as those in $(1 + x)^2$.

To discover the Binomial Theorem one need only work out some powers of the type indicated above, tabulate the resulting data in an orderly way, and look for formulas behind the numbers. It is a straightforward exercise on the approach explained in the previous chapter, 'A Method of Discovery'.

In this chapter, as in the last, the signs * * * will be provided for the reader who wishes to make his own investigation before reading the explanation. It is strongly recommended that this path should be followed. The experience of discovery is much more interesting than passive learning. It increases confidence greatly, gives practice in using the procedures of algebra, and helps the memory; you are much more likely to remember something you have thought out for yourself than something you have just been told. Further, if after some lapse of time you should forget the result, you know you can work it out again for yourself. The best way to learn mathematics is by repeatedly thinking it out.

	1	x	x^2	x^3	x^4	x^5	x^6
$n = 0$							
$n = 1$							
$n = 2$	1	2	1				
$n = 3$	1	3	3	1			
$n = 4$							
$n = 5$							
$n = 6$							

Figure 74

The Binomial Theorem is about expressions of the form $(x + y)^n$ or $(1 + x)^n$. Obviously our first step is to collect data on such expressions. The same numbers appear in each, so we will work with the simpler expression, $(1 + x)^n$. Let us work this out for $n = 0, 1, 2, 3, 4, 5, 6$, and tabulate the results in the form shown in Figure 74. The rows correspond to the various values of n. Thus, opposite $n = 2$ we see the numbers 1, 2, 1 that occur in $(1 + x)^2$. These are entered in the columns headed 1, x, and x^2 because the full expression for $(1 + x)^2$ is $1 + 2x + x^2$. We have already worked out $(1 + x)^3$ and the numbers 1, 3, 3, 1 have been entered in the row for $n = 3$. The remaining rows need to be filled in.

*   *   *

If we multiply our result for $(1 + x)^3$ by $1 + x$, we find $(1 + x)^4 = 1 + 4x + 6x^2 + 4x^3 + x^4$; accordingly we enter the numbers 1, 4, 6, 4, 1 in the row for $n = 4$. Similarly for $n = 5$, we find the numbers 1, 5, 10, 10, 5, 1, and for $n = 6$ we find 1, 6, 15, 20, 15, 6, 1. At the top we still have to fill in the rows for $n = 1$ and $n = 0$. For $n = 1$, we have simply $1 + x$, so we write in 1, 1. The row with $n = 0$ will trouble some people, On page 88, we saw that for every number a, the value of a^0 is 1. So $(1 + x)^0 = 1$. After entering all this information, we have a table that looks like this:

	1	x	x^2	x^3	x^4	x^5	x^6
$n = 0$	1						
$n = 1$	1	1					
$n = 2$	1	2	1				
$n = 3$	1	3	3	1			
$n = 4$	1	4	6	4	1		
$n = 5$	1	5	10	10	5	1	
$n = 6$	1	6	15	20	15	6	1

Figure 75

Discovering the Binomial Theorem

One point remains to be discussed. There are still some blank spaces. Should these spaces be left blank? Would it be stupid to try to fill them in, like looking for the answer to a meaningless question? Or is there a definite number that ought to be written in each of these spaces?

*   *   *

To settle this point, we must think what our numbers mean. In the row $n = 6$ we see 15 in the x^4 column because $15x^4$ occurs in $(1 + x)^6$. In formal language, 15 is the coefficient of x^4 in $(1 + x)^6$. In rather informal language, it tells us how many times x^4 appears in $(1 + x)^6$. In the same column we see a blank in the row $n = 2$. To write a number in this space, we would have to answer the question, 'How many times does x^4 appear in $(1 + x)^2$?' There is a perfectly good answer to this question, – 'None'. We usually write $(1 + x)^2$ simply as $1 + 2x + x^2$. But it would still be perfectly correct if we wrote $1 + 2x + x^2 + 0x^3 + 0x^4 + 0x^5 + 0x^6$. Accordingly the coefficient of x^4 in $(1 + x)^2$ is 0 (The same argument applies to all the other empty spaces. In each of them we can correctly write a nought. Our table is now complete and appears as follows:

	1	x	x^2	x^3	x^4	x^5	x^6
$n = 0$	1	0	0	0	0	0	0
$n = 1$	1	1	0	0	0	0	0
$n = 2$	1	2	1	0	0	0	0
$n = 3$	1	3	3	1	0	0	0
$n = 4$	1	4	6	4	1	0	0
$n = 5$	1	5	10	10	5	1	0
$n = 6$	1	6	15	20	15	6	1

Figure 76

Now we have enough evidence and can start looking for regularities. You will see immediately that there is a simple pattern in the first two columns. For the remaining columns you will need to use the methods of the previous chapter. A pattern runs through the whole thing, and this appears most clearly when the formula for each column is left in such a form that its factors are seen. You will lose the pattern if you multiply out.

*   *   *

Evidently the first column always contains 1. In the x-column, we see 2 in the row $n = 2$ and 4 in the row $n = 4$. In fact, the number in this column is always equal to n. Thus in $(1 + x)^n$, the constant term is always 1 and the coefficient of x is n. This means that $(1 + x)^n$ begins $1 + nx + \ldots$.

In the x^2 column we see the numbers 0, 0, 1, 3, 6, 10, 15. We apply to these the procedure of the 'trick' in our last chapter, and get the table

0		0		1		3		6		10		15
	0		1		2		3		4		5	
		1		1		1		1		1		

There are two ways of making our next step. If we apply the procedure of the original trick, we are led to the formula $\frac{1}{2}n^2 - \frac{1}{2}n$. (Note that it is n that comes here, not x, for these numbers correspond to the rows $n = 0, n = 1, n = 2$, and so on. We are using x, x^2, and x^3 to label columns. It would cause great confusion if we used x also to give the rule behind these numbers.)

But we can also use equation (i) of page 187. We could write another row of differences in the table we have just used. It would then look like the table for a cubic, but with noughts in the bottom row.

0		0		1		3		6		10		15
	0		1		2		3		4		5	
		1		1		1		1		1		
			0		**0**		**0**		**0**			

The numbers in heavy type indicate that here $p = 0$, $q = 0$, $r = 1$, $s = 0$. Substituting these in equation (i), and being careful to put n instead of x we find that $\dfrac{n(n-1)}{1.2}$ gives the numbers in this column.

It is not really necessary to write the extra row in the table. Any quadratic gives noughts in the third row of differences, so always $s = 0$. For a quadratic, we can simply leave out the term involving s in equation (i).

Of course the two methods used above lead to the same answer. The first method gives it multiplied out; the second method gives it in factors.

The numbers in the x^3 column lead us to the table

$$
\begin{array}{ccccccc}
0 & 0 & 0 & 1 & 4 & 10 & 20 \\
 & 0 & 0 & 1 & 3 & 6 & 10 \\
 & & 0 & 1 & 2 & 3 & 4 \\
 & & & 1 & 1 & 1 & 1
\end{array}
$$

This indicates a cubic, with $p = 0$, $q = 0$, $r = 0$, $s = 1$ and equation (i) gives us the formula $\dfrac{n(n-1)(n-2)}{1.2.3}$. The same thing happens in the columns headed x^4, x^5, and x^6. In each case we find that all the numbers along the sloping edge of the table, with one exception, are 0. The remaining number is 1. Thus an equation, having the same pattern as equation (i) but with more terms, gives the formula. So many of the numbers $p, q, r, s, \ldots$ are 0 that it is quite easy to make the substitution. Only one term appears in the result, and in fact you can quite easily guess what that will be by looking at the results for the x^2 and x^3 columns.

It was mentioned on page 188 that expressions such as $\dfrac{n(n-1)}{1.2}$ and $\dfrac{n(n-1)(n-2)}{1.2.3}$ are known as the Binomial Coefficients. We now have the reason for this name. These expressions give the coefficients in $(x + y)^n$ and $(1 + x)^n$. The expressions $x + y$ and $1 + x$ are described as *binomials* because they contain two terms. In the same way, $5x + 2y - 3z$ is sometimes described as a trinomial because it contains the three parts, $5x$, $2y$, and $3z$, linked together by $+$ and $-$ signs. However, it is not particularly important to know these names. The important thing is to understand the algebra itself, not the names which various writers have found it convenient to coin. You could be a master of algebra without knowing any of these names. However, for what it is worth, the Binomial Theorem is so called because it enables us to write down the powers of expressions, such as $1 + x$ or $x + y$, which consist of two parts. The Binomial Coefficients are the coefficients found in these powers. If we collect together the formulas we have found for the various coefficients, we arrive at the following result

$$(1 + x)^n = 1 + \frac{n}{1}x + \frac{n(n-1)}{1.2}x^2 + \frac{n(n-1)(n-2)}{1.2.3}x^3 + \ldots$$

It is understood that further terms appear in the place where we see a row of dots. These continue the pattern, which will be easier for you to see than for me to explain in words. The result above is the Binomial Theorem.

The dots at the end suggest that it goes on for ever, whereas an expression like $(1 + x)^2$ stops after three terms, $1 + 2x + x^2$. However, you may remember we put noughts in the blank spaces of the table, and the formulas we found were chosen to give these. So in fact the expression given above automatically turns itself off at the right place. It does this by producing noughts. For instance, if we put $n = 2$ in the equation for $(1 + x)^n$ we have just had, we find $(1 + x)^2 = 1 + 2x + x^2 + 0x^3 + \ldots$.

AN ALTERNATIVE APPROACH

It is possible to see why factors such as $n - 1$ and $n - 2$ *must* appear in the binomial coefficients, and it is the noughts in the table that give us the clue. Let us consider again the numbers in the x^3 column. x^3 does not occur in 1, in $1 + x$ or in $(1 + x)^2$, so the first three numbers are bound to be 0, 0, 0. Now we saw when we were considering the Remainder Theorem that if the polynomial expression $f(n)$ takes the value 0 when $n = a$, then $n - a$ must be a factor of $f(n)$. Now in the x^3 column we find the value 0 for $n = 0$, $n = 1$ and $n = 2$. It follows that the expression behind these numbers must have the factors $n - 2$, $n - 1$, and $n - 0$. The last of these may be written more shortly as simply n.

Accordingly $n(n-1)(n-2)$ must divide exactly into this expression; the expression must be of the form '$n(n-1)(n-2)$ times something'. What can the something be? We found earlier, by counting the rows of differences, that a cubic expression was to be expected. Now $n(n-1)(n-2)$ is already a cubic; it equals $n^3 - 3n^2 + 2n$. This limits very much what the 'something' can be. Suppose, for example, someone suggested the 'something' might be $2n^2 + 3n + 4$. Calculating $n^3 - 3n^2 + 2n$ times $2n^2 + 3n + 4$ we would obtain an expression containing $2n^5$. But this cannot be right; the rows of differences told us not to expect any power higher than n^3. We try again. How would $5n + 1$ do for the 'something'? It is no better, since $n^3 - 3n^2 + 2n$ times $5n + 1$ contains $5n^4$, which is still too high a power to be acceptable. We are forced to the conclusion that the 'something' must be simply a number. What number we cannot say on the evidence of 0, 0, 0 for $n = 1, 2, 3$. Whether we take $7n(n-1)(n-2)$ or $100n(n-1)(n-2)$ or $\frac{1}{2}n(n-1)(n-2)$, when we put $n = 0, 1, 2$ in turn, we get 0, 0, 0. All we can say – for the moment – is that the answer must be $kn(n-1)(n-2)$ where k stands for *some* number. To fix k we must look at the other numbers in the x^3 column. Now the numbers in the x^3 column read 0, 0, 0, 1, 4, 10, 20. This means we have to choose k so that $kn(n-1)(n-2)$ gives 1 for $n = 3$, 4 for $n = 4$, 10 for $n = 5$, and 20 for $n = 6$. Here are four conditions to be satisfied by the single number k. We cannot force one unknown to satisfy four equations. Any single equation will fix k; we can only hope (or expect on theoretical grounds) that the same value of k will satisfy the other three equations. Let us then take the first condition; putting $n = 3$ is to give the value 1. Putting $n = 3$ in $kn(n-1)(n-2)$ we get $k.3.2.1$ or $6k$. Accordingly our equation for k is $6k = 1$, and so $k = \frac{1}{6}$. So the expression that gives the numbers in the x^3 column is $(1/6)n(n-1)(n-2)$.

Actually, we would have been wiser not to have multiplied $k.3.2.1$ out as $6k$, for in its first form it gives us not merely the number 6, but the pattern $3.2.1$ of the number that occurs in the denominator of $\dfrac{n(n-1)(n-2)}{1.2.3}$. For when we put $n = 3$ in this we find on top $3.2.1$. If we are to get 1 when we put $n = 3$ in the whole expression, we must have $3.2.1$ down below to balance the 3.2.1 on top. And that is exactly what we do have. (Whether we write the denominator as 3.2.1 or 1.2.3 is purely a question of taste. Both represent the same number.)

USES OF THE BINOMIAL THEOREM

The Binomial Theorem, as we saw earlier, is concerned with a formula for the numbers that occur when we multiply out an expression of the form $(1 + x)^n$. This immediately raises the question – what use is such a theorem? We do not often have to deal with a power higher than the sixth, and the table in Figure 75 gives us the numbers involved in all the powers up to $(1 + x)^6$. Why should we bother about a formula? Why not just read the numbers from the table?

In the chapter 'Making the Curved Straight' it was mentioned that Wallis and Newton developed the idea of negative and fractional indices because they needed these for their work on calculus. This work on calculus frequently involved expressions

such as $\sqrt{(1 + x)}$, $\quad \dfrac{1}{\sqrt{(1 + x)}}$, $\quad \dfrac{1}{1 + x}$, $\quad$ or $\quad \dfrac{1}{(1 + x)^2}$.

Now these expressions, if we use fractional and negative indices, can be written as $(1 + x)^{\frac{1}{2}}$, $(1 + x)^{-\frac{1}{2}}$, $(1 + x)^{-1}$, and $(1 + x)^{-2}$; that is, they are all of the form $(1 + x)^n$. In the early part of this chapter we were led to guess the Binomial Theorem by observing what happened in the cases where n was 0, 1, 2, 3, 4, 5, or 6. So all our evidence was based on whole number values for n. Newton, faced with awkward expressions such as those listed above, made another guess. How nice it would be if the Binomial Theorem worked not only for whole numbers, but also for fractions and negative numbers. He tested this idea and found that he could get correct results by using it. For instance, to deal with $\sqrt{(1 + x)}$, which is $(1 + x)^{\frac{1}{2}}$, he would take the Binomial Theorem and substitute $n = \frac{1}{2}$. The Binomial Theorem is

$$(1 + x)^n = 1 + \frac{n}{1}x + \frac{n(n - 1)}{1.2}x^2 + \frac{n(n - 1)(n - 2)}{1.2.3}x^3 + \cdots$$

If we put $n = \frac{1}{2}$, then $n - 1$ becomes $-\frac{1}{2}$, $n - 2$ becomes $-\frac{3}{2}$, and the later terms, indicated by dots, continue the pattern. Accordingly $n(n - 1)/(1.2)$, the coefficient of x^2, becomes $(\frac{1}{2})(-\frac{1}{2})/(1.2)$, which works out as $-\frac{1}{8}$. In the same way the coefficient of x^3 becomes $(\frac{1}{2})(-\frac{1}{2})(-\frac{3}{2})/(1.2.3)$, which boils down to $\frac{1}{16}$. If we need the terms containing x^4 and higher powers, we can work them out too. We thus arrive at the result

$$(1 + x)^{\frac{1}{2}} = 1 + \frac{x}{2} - \frac{x^2}{8} + \frac{x^3}{16} \cdots$$

It should be emphasized that we have only guessed this result. The expression on the right-hand side goes on for ever, unlike the whole number cases, in which the terms automatically became zero after a certain point. Infinite series are tricky things that always have to be handled with caution. It has been found that our series for $(1 + x)^{\frac{1}{2}}$ can only be used safely for values of x between -1 and $+1$. It is misleading if used outside this interval. It is most helpful when x is very small. For instance, by putting $x = 0.01$, and using only the terms actually printed above (that is, ignoring x^4 and all higher powers), you can find $\sqrt{1.01}$ correct to eight places of decimals. The work is not difficult and this method compares favourably with any other way of calculating this number to such a high degree of accuracy.

Actually, this is a comparatively unimportant application of the Binomial Theorem, but it may perhaps show how a door is opened on all kinds of possibilities. On the left-hand side of our last equation we see $(1 + x)^{\frac{1}{2}}$, which means $\sqrt{(1 + x)}$; a square root is a slightly awkward thing to handle or to work out. On the right-hand side we see (if we ignore the dots) a cubic. This is a very simple expression, easy to handle, and in fact very much the kind of expression that a beginner meets at the start of an algebra book – a polynomial. This polynomial is sufficiently close to the square root, over a certain interval of values of x, to allow us to replace one by the other. Here is an idea and a method of great scope and power – that an awkward, complicated expression can be fitted quite closely by a simple polynomial. The development of this idea is found in books on calculus, in the chapters dealing with the topic of series.

THE BINOMIAL COEFFICIENTS

The numbers we met in Figure 75 can with advantage be arranged in the form of a triangle, as here:

```
            1
          1   1
        1   2   1
      1   3   3   1
    1   4   6   4   1
  1   5  10  10   5   1
1   6  15  20  15   6   1
```

There is a simple rule for writing these numbers. Along two edges we see the number 1 repeated. Each number inside the triangle can be obtained by adding together the two numbers immediately above it. These numbers have a way of popping up in different parts of mathematics, and it is well worth while to write out the numbers above (using the rule just explained) from time to time, until these numbers become quite familiar and are immediately recognized if they occur in any piece of work.

The array of numbers above is usually called Pascal's Triangle, and Pascal did indeed work on this diagram in 1653. However, according to an article by A. N. Singh in the first volume of the historical journal *Osiris* (pages 623–4), this triangle was known to Pingala in India around 200 B.C. and arose in a question relating to poetry. Rhythm depends on the use of long and short syllables, denoted by — and ∪ respectively. The question is, to classify the different metres that can be produced with any number, n, of syllables. If only one syllable is used, there are of course only two possibilities, it may be long or short. With two syllables, we have the four possibilities

$$— —; \quad — ∪; \quad ∪ —; \quad ∪ ∪$$

Pingala grouped the two middle cases together, since in each of them we see 1 long and 1 short. Thus the four possibilities are grouped into classes with 1, 2, 1 members. With three syllables there are eight possibilities. Only one of these, — — —, has all the syllables long. There are three possibilities with two long and one short, namely — — ∪; — ∪ —; ∪ — —. Similarly there are three with one long and two short, and finally one with only short syllables. Thus eight is broken up into 1, 3, 3, 1. The enumeration continues. At the next step we find sixteen cases, which break up to reveal the pattern 1, 4, 6, 4, 1. It will be seen that the numbers arising here are the same as those in the rows of the triangle at the beginning of this section. Pingala arranged his numbers in a triangle; he was aware that each number could be obtained by adding together the numbers above it; he was also aware that adding up the numbers in each row gave the powers of 2. Thus $1 + 1 = 2, 1 + 2 + 1 = 4, 1 + 3 + 3 + 1 = 8, 1 + 4 + 6 + 4 + 1 = 16$, and so on.

This ancient investigation of metres is closely related to one of the ways in which the binomial coefficients are being actively used today. Essentially, Pingala was investigating the number of ways in which a certain thing could happen. For instance, he found that there were two ways, $— \cup$ and $\cup —$, of getting a two-syllable sound with one long and one short. Now in poetry I do not suppose that two-syllable words or two-syllable feet of these two types occur equally often. The iambic foot, di-dah, must surely occur more often than the trochee, dah-di. But there are many situations in which the likelihood of something happening depends on the number of ways in which it can happen. Suppose, for instance, you spin two coins, a silver coin and a copper coin. Each coin, we suppose, is equally likely to come down heads or tails. If we write a capital letter for the result with the silver coin, and a small letter for the copper coin, there are four possibilities, all equally likely, — Hh, Ht, Th, Tt. Of these four, only one has both coins heads, namely Hh, and only one has both tails, namely Tt. However there are two cases in which we get a head and a tail, Ht and Th. This outcome is therefore twice as likely as getting a pair of heads. If we threw a couple of coins a thousand times, we would expect to get two heads *about* 250 times, a head and a tail about 500 times, and two tails about 250 times. The word *about* is emphasized, since chance variations occur. If you tossed a pair of coins ten times you would expect to get a pair of heads round about two or three times. But there is something like one chance in a million that all the coins would come down heads, and this one time might be when you did your experiment.

Except for professional gamblers the odds on penny tossing are not particularly important. But there is a tremendous gamble in which we are all involved, the process by which life has appeared and is carried on. The heredity of each individual seems to depend on chance recombinations of the genes carried by his parents, and the appearance of new species on mutations, random and unplanned changes in the chemistry of genes. Through mutations, progress is bought at the price of suffering. The great majority of chance variations will be harmful, but once in a while the new gene structure will be more efficient than the old; the mutant will thrive and his kind will multiply. It is this selection of successful innovations which gives the apparent order to living things, and makes it possible for such complex and organized things as plants and animals to emerge. The variations appear by chance; the creatures that survive embody a whole series of variations in a combination that has been tested by the dangers and difficulties of millions of years of existence. One of the most interesting ideas in recent genetics is that animals carry within themselves not merely defences against present dangers, but also reserves of variations to meet dangers that have not yet arisen. The flies had no way of knowing that men were going to invent D.D.T., but within a short time after its widespread use, new strains of insects resistant to D.D.T. became established. No doubt there are limits to this. I do not suppose that we carry within our genetic reserves variations which would enable us to live at the temperature of a coal fire.

Near the beginning of this chapter, it was mentioned that the ratios 1:2:1 appear in the mating of roan cattle, the same ratios we have just met in the tossing of a couple of coins. And the mechanism is indeed very similar. The bull carries within himself the red and the white strains, and is equally likely to give either to one of his offspring. This corresponds to the tossing of the first coin. The cow also is equally likely to hand on a red or a white unit; this is the tossing of the second coin. If we write R for red and W for white, the four equally likely inheritances for the calf are RR, RW, WR, and WW. The first of these makes the calf pure-bred red, he will look red and have nothing but red inheritance to hand on to his eventual offspring. The next two cases, RW and WR, make the calf both look like his roan parents and be like them in the heredity he can hand on.

It should perhaps be mentioned that the situation here, in which the genetic make-up of an animal can be recognized from its appearance, is by no means the usual one. There are many cases in which an animal carries a gene the existence of which cannot be deduced from the animal's external appearance.

The penny-tossing situation is also closely related to the question of how many boys and girls one should expect in a family. The chance of a baby being a boy or a girl is not exactly even. Women live longer than men, but this is offset by more boys being born than girls. In the Canada Year Book for 1963–4, the section on vital statistics contains the following remarks, 'Wherever birth statistics have been collected they have shown an excess of male over female births. No conclusive explanation of this excess has yet been given. ... The number of males to every 1,000 females born in Canada has averaged around 1,057 since the middle 1930s.' If we merely want to get a rough idea of how the probabilities work out we may ignore this difference. Let us suppose then, for simplicity, that the chances of getting a boy or a girl are just like the chances of getting a head or a tail on a spun penny. Then, if two children are born in a family, the four cases BB, BG, GB, GG are equally likely. (Here, as may well be guessed, BG indicates that the first child is a boy, the second a girl.) The familiar 1:2:1 appears again in the chances of getting both boys, a boy and a girl (in either order), or both girls. If a family has four children, there are sixteen possible cases, if we take account of the order in which the children are born, going from (1) BBBB, (2) BBBG, to (15) GGGB, (16) GGGG when the cases are arranged in alphabetical order. But if we are only concerned with how many boys there are and how many girls, we are dealing with exactly the same question as Pingala when he classified the possible metres of four-syllable words. We shall arrive at the same numbers 1, 4, 6, 4, 1; there is one case with four boys while there are six cases with two boys and two girls.

The same numbers appear in the question of how people's heights vary. It seems that the height of a person (to the extent that it is determined by heredity, and not by nutrition or other

203

circumstances) depends on several genes. If all the genes are working together to produce tallness, a very tall person results; if they all work for shortness, a very short person results; if, as is most likely in someone selected at random from the population, some of the genes are working for tallness and some for shortness, a person not very far from average height should result. We can make a model to show the *kind* of process involved (it makes no claim to reflect the latest research on the inheritance of height) by imagining height to depend on the tossing of four coins. Suppose we mark +1 on each head, and −1 on each tail, with the

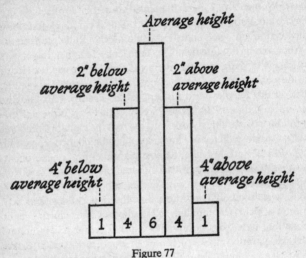

Figure 77

understanding that +1 indicates a contribution tending to make a person one inch above average height, and −1 indicating one inch below average height. The actual height of the person is to be found by adding all the contributions. Thus throwing four heads would give +1 + 1 + 1 + 1 and make the person four inches above average height. Throwing two heads and two tails would give +1 + 1 − 1 − 1 (in some order); as the total is 0, this would give average height. This is the most likely case, occurring six times out of sixteen. The distribution of heights that would result from this procedure is shown in Figure 77.

Obviously, this is a rather crude model. If the average height of a man is 5 ft 7 in. this model would predict that all men had one of the heights 5 ft 3 in., 5 ft 5 in., 5 ft 7 in., 5 ft 9 in., 5 ft 11 in. Nevertheless, the diagram obtained has a general resemblance to the way in which heights actually are distributed, with heights near the average being very common, and heights much above or below the average quite rare. If we want to get closer to the actual distribution of heights we may have to consider a genetic mechanism which corresponds to the tossing of more than four coins. It may also well be that the height of a man depends on factors other than genes, such as the health of his mother in the months before he was born and the general conditions of the environment in which he grew up. It is possible to produce more elaborate models in which all these chances are simulated by the tossing of coins and the throwing of dice (some of them loaded) to determine whether the individual is to be born into a family rich, middling, or poor; underfed, well-fed, or over-fed; athletic, active, or lazy; teetotal, alcoholic, or drug addicted; smoking or non-smoking – and so forth indefinitely. The more coins are tossed in the model, and the smaller the contribution that each coin makes to the final result, the smoother will be the curve of the predicted distribution. Much work in biological and social investigations is done by considering various simplified mechanisms, akin to the coin tossing just described, and seeing which mechanism comes closest to predicting the effects actually observed. One distribution, the 'normal curve', corresponds to throwing a very large number of coins, each of which makes a very small contribution. This gives a bell-shaped curve, and it was at one time believed that some fundamental simplicity in the nature of things required *all* statistical phenomena to conform to this curve. This is far from being so; this curve is one of the possibilities that have to be considered in any inquiry. It can be shown, though this requires a little more calculus than many users of statistics have, that the shape of the curve is essentially given by the equation $y = e^{-x^2}$. The symbol e occurring here is the number that occurs at distance 1 when a logarithmic scale is constructed in the way that Napier used. We referred to this briefly on page 82. Anyone working on statistics will keep meeting this symbol e, and will gain a sense of security by knowing what e

is and what it does. Some of the older books, such as Hall and Knight's *Elementary Algebra* and *Higher Algebra* give a treatment of e which does not assume any knowledge of calculus. This treatment is painful in the extreme, and should be avoided. In *Mathematician's Delight* I tried to show that, by using the seventeenth-century approach to logarithms and some simple ideas about calculus, it was possible both to picture the meaning of e and to have at any rate an informal understanding of its main mathematical properties. It would be wise for anyone intending to work with statistics to get at least this amount of knowledge about e and the basic ideas of calculus. It is most disturbing continually to be using formulas containing a symbol, the meaning of which you do not understand.

In the past, calculus used to be a word of terror. An interesting comment on the change in this respect is to be found on page 20 of the Ministry of Education pamphlet, *Teaching Mathematics in Secondary Schools*. In 1944, it was suggested that an alternative syllabus be made available to the younger pupils in secondary schools, with less heavy work in algebra, but with some calculus brought in. The pamphlet remarks that some schools (not all, I admit) regard this new option as most suitable for *their weaker pupils*. The theory of e is an excellent illustration of this view. A large proportion of pupils can understand the basic ideas of calculus and can appreciate the power and simplicity with which calculus arrives at the properties of e. The treatment of these properties by algebra would require a capacity for sustained concentration and a courage in dealing with long and complicated expressions that relatively few people have.

Incidentally, while speaking of long and complicated expressions, we may observe that the relatively simple $y = e^{-x^2}$ has a way of turning up in statistical work in forms that effectively camouflage its simplicity.

Three modifications of it may occur. As it stands, it gives a bell-shaped graph with a maximum at $x = 0$. But, of course, in many situations the average value, the most frequent value, is not 0. For example, the average height of a man is not nothing. Now the bell shape is concerned with deviations from the average. So if x measures the height of a man, and a represents the average height of a man, the equation will involve $x - a$, the deviation of the man's height from the average. We shall be concerned not with $y = e^{-x^2}$ but with $y = e^{-(x-a)^2}$. This represents exactly the same bell-shaped curve, but it has been slid along the x-axis until its maximum occurs at $x = a$.

Also bells differ in breadth. If an excellent marksman is shooting at a target placed at the point $x = 0$, all his shots may lie extremely close to it. A poor marksman may have many shots at a considerable distance. The distribution curves for two riflemen may be as in Figure 78. The curve in (A) was drawn from the equation $y = e^{-x^2}$, that in (B) from $y = e^{-4x^2}$. Both curves have the bell shape, but (A) is more spread out than (B). Either curve could be obtained from the other by changing the scale on the x-axis while leaving the scale on the y-axis unaltered. An attempt is made to show this pictorially. In the left-hand half of each curve a number of upright lines are drawn. If we think of these as upright posts, then in (A) and (B) we see the same collection of posts, but the spacing between them is only half as much in (B) as in (A). This comes out clearly when you plot the graphs. In (A), putting $x = 1$ in e^{-x^2} gives e^{-1}, which a table shows to be (approximately) 0·37. In (B), putting $x = \frac{1}{2}$ also gives e^{-1}, that is, 0·37.

Quite generally, such changes of horizontal scale correspond to taking different values for k in e^{-kx^2}.

Thirdly, we can get changes in vertical scale by taking $y = ce^{-x^2}$ instead of $y = e^{-x^2}$.

Usually, all three effects are combined, and we have to deal with $y = ce^{-k(x-a)^2}$. It is important to realize that this equation does not represent anything more complicated than the original $y = e^{-x^2}$. It is quite easy to imagine a bell curve being slid horizontally, and being stretched or compressed horizontally and vertically.

Finally, the numbers c, k, and a are likely to have some history. They have been arrived at by some computation, practical or theoretical. For instance, on page 109 of Moroney's book, *Facts from Figures* (Penguin Books), c takes the form

$\dfrac{1}{\sigma\sqrt{2\pi}}$, k is $\dfrac{1}{2\sigma^2}$ and a is $\bar{x}$. However, in any particular example,

when numerical values are put in, these expressions will give

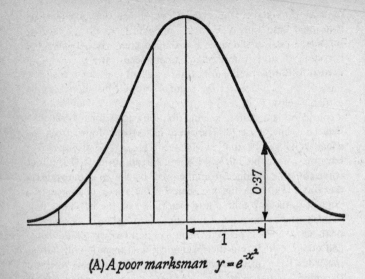

(A) *A poor marksman* $y = e^{-x^2}$

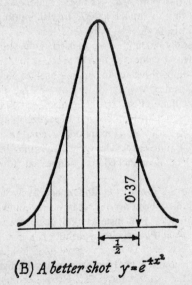

(B) *A better shot* $y = e^{-4x^2}$

Figure 78

us a single number for each of c, k, and a. As has been mentioned before, in algebra one should always be aware that formulas look worse than they really are because algebra retains a record of how a number was arrived at, while arithmetic obliterates this record.

IN HOW MANY WAYS?

Chance pervades life. We have already mentioned its role in heredity. The theory of chance plays a great role in physics and chemistry, where so many millions of atoms are involved that we cannot follow their individual careers, but find that we get quite an effective theory of, say, the behaviour of a gas by treating the collisions of the molecules as purely random events. This approach constitutes the method of Statistical Mechanics. Statistical methods are important in many biological and social investigations. A statistical test of experimental data says, in effect, 'Either we have here evidence for the truth of a certain belief or else some very rare and unusual accident has occurred, by pure chance.' Problems of statistics and chance occur frequently in industry. If, most days, 1,000 telephone calls are made between noon and 1 p.m., each lasting on the average 5 minutes, how many telephone wires should be provided to make it unlikely that any customer will have to wait more than 30 seconds for a free line? The word *unlikely* should be noticed. In exceptional circumstances, it might happen that all the 1,000 callers would reach for the phone at precisely 12.57 p.m., but this should happen so rarely that the officials responsible for planning telephone services should survive the criticism. It would be uneconomic to provide 1,000 telephone lines on the grounds that this might happen.

Again industry uses statistics for the purpose of quality control. It is always possible that some machine may go out of order in an unobtrusive way. A manufacturer does not want to flood the market with defective articles bearing his name. But, on the other hand, it is expensive and sometimes impossible to test every article. A very hackneyed example is the manufacture of cartridges. If the manufacturer tests every cartridge by firing it, he will have none to sell. His aim is to test enough cartridges to make it unlikely that a defective batch will slip through unnoticed. Here again, he can only make it *unlikely*; nothing can make it *impossible*. But 'unlikely' is good enough. If defective goods go out only rarely, the company can afford to type out letters of profuse apology and send out free replacements and perhaps compensation for loss. How unlikely a defective cartridge must be made will of course depend on the risk to which the user is exposed. The level could be lower for a hunter of rabbits than for a hunter of rhinoceros.

Nearly all problems about probability and chance can be simulated by problems about spinning coins, throwing dice, or drawing cards from a pack. This is convenient. It eases thought. In working a problem of heredity we do not need all the time to picture the complicated processes of meiosis and mitosis. We can speak (as was done earlier in this chapter) of the spinning of a coin as deciding whether the calf receives the gene for redness or whiteness from its father. We must of course check that we are not oversimplifying when we use such a device. There is also the advantage that, by speaking in terms of spinning coins and so forth, we can teach the mathematics of probability to a class in which some learners are interested in heredity, some in sociology, some in industry, and some in physics.

Some care is needed in experiments intended to illustrate statistical ideas. One difficulty is that the unusual can occur. There is a story about a man in the early days of quality control who gathered an audience of millionaires around him, to explain how this new technique would increase the profits of their factories. He did a demonstration to illustrate the principles involved. He had six jars, each containing nine white balls and one black ball. He proposed to draw one ball from each jar, and pointed out that the chance of getting all black balls was only one in a million; this, he said, was so small that 'in practice we can neglect it'. He then drew out six balls – and each one was black! What he should have said, of course, was that a procedure that fails only once in a million times is good enough for many practical purposes.

Another difficulty is to make selection really random. It is very difficult to shuffle a pack of cards so thoroughly as to destroy all traces of the previous order of the pack.

Discovering the Binomial Theorem

Most problems of probability involve problems of counting. What is the chance that a family of four children will consist of two boys and two girls? We considered this earlier. First we had to establish that there were sixteen possible cases for a family of four children and check (or agree to accept) that each case was equally likely. Thus the situation is the same as if we had sixteen cards, the first marked BBBB, the next BBBG, and so on to the last GGGG. We then have to pick one of these at random. What is the chance we shall get a card with B on it twice and G on it twice? To answer, we must find out how many such cards there are. We find there are six of them, so the chance is six out of sixteen.

Similarly, if we are asked, 'When four cards are dealt from a pack of fifty-two cards, what is the chance that all four are aces?' our job is to count how many ways there are of selecting four cards from fifty-two, and how many of these selections will consist of four aces. If you know nothing about football, and are asked to predict, for each of seven teams, whether they will win, draw, or lose their next match, what is your chance of being correct? Here again, we have to find how many possible outcomes there are in seven games. Since you will be right in only one case, your chance is one in that number.

Such questions are considered in algebra books, in a chapter headed 'Permutations and Combinations'. Such chapters usually have a rather bad reputation with learners. One important (and avoidable) reason is that the work is taken far too fast. Learners become used to the idea that in algebra there are certain routine methods; you learn these and apply them. No one can learn permutations and combinations in this spirit. Rather, each question must be approached as a problem. The problem sounds simple – it is just one of counting. But the possibilities involved are so many that the detailed counting of them would be wearisome or impossible in any reasonable time. One therefore has the problem of listing the possibilities in some systematic way, to make sure that none is overlooked and none is counted twice. In a really complicated problem, this may be very difficult indeed. There are problems of this kind, of interest in science, that are still unsolved. But there are many relatively straightforward cases, which one can learn to handle and which in time become familiar.

Not only are learners often in too much of a hurry to reach an answer. Often the writers of books assume that after a short paragraph of explanation the whole principle will be clear to the reader. This is a most questionable assumption. Generally speaking, it is necessary for a learner to take a question and play around with it and make experiments until the meaning gradually sinks in. You may have noticed that questions 3 and 6 at the end of the chapter 'A Method of Discovery' were in fact problems of this kind, which the reader was asked to explore for himself, before any mention had been made of permutations and combinations. It is not difficult to work out how many tickets are needed on a railway with four stations. We simply have to take reasonable care not to leave any possibilities out, and this already gets us thinking about the central question – how to arrange the possibilities systematically, to make sure that none gets overlooked. Of course, we could not follow the procedure of simply enumerating all possibilities if the question were, 'How many tickets for a line with 279 stations?' Our experiments have to be with questions where small numbers are involved. Having collected data in this way, we may examine our results and observe that some formula fits them, as in the chapter 'A Method of Discovery'. Or it may be that in arranging our work, we observe a principle that could be applied to large numbers as well as to small ones. For instance, we might reasonably arrange the tickets for the four-station railway in this manner:

AB	AC	AD
BA	BC	BD
CA	CB	CD
DA	DB	DC

In the first line, we have all the tickets that start from station A; they must go to B or C or D. In the next line we have all the tickets starting from station B, and so on. We can hardly fail to see the shape of a rectangle here; in *Vision* it was emphasized that the rectangle illustrates multiplication, and our arrangement above suggests 4 times 3. If there were n stations, we would have n rows, one for each station from which a traveller could depart. How many different tickets would be sold at that station? The traveller may go to any other station on the line, and there are

$n-1$ of these. Thus, in general, we would have n rows with $n-1$ in each, and the total number of tickets would be $n(n-1)$. In this way, we have been led from our work with four stations not merely to guess the answer for n stations but to prove by a reasoned argument what that answer must be.

Both procedures discussed above should be followed freely – guessing the general formula from particular data and finding a method of organizing the possibilities that works in the general case.

Often the first method, guessing the answer, helps us to construct a proof. For instance, in the question of how many diagonals an n-sided figure has we arrived at the formula $\frac{1}{2}n(n-3)$. Thus, for example, the seven-sided figure has $\frac{1}{2} \times 7 \times 4$ diagonals. Why do we get this answer? For instance, why should the factor 7 be there? There are seven corners from which a diagonal can be drawn, so it is natural that 7 should be involved. As Figure 79 shows, four diagonals sprout from each corner. This

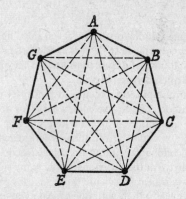

Figure 79

number 4, as the factor $n-3$ indicates, is there because it is 3 less than 7. Why 3 less? (In a classroom, of course, the pupils should find the answer. In a book, I can only give the answer – though, as usual, the reader may pause.)

* * *

The reason is, that from any corner, say D, we get a diagonal by joining D to any corner except itself or its neighbours, C and E.

Finally, why should there be a factor $\frac{1}{2}$ in the formula? Why is the number of diagonals not simply 7×4, since four diagonals sprout from each of seven corners?

If we go round the corners, inking in each diagonal that springs from that corner, we soon notice that this causes us to count each diagonal twice; the diagonal FB is inked once when we are at corner B and gets inked again (if we persist in the procedure) when we come to the corner F.

It is now easy to extract from the considerations above a proof of the formula. In many books this proof is given. Many learners will find the proof far more meaningful if they have arrived at it by a route such as that just explained. A proof means nothing so long as it is simply on paper. Only in the thinking of the learner does it come to life.

CODES AND SIGNALS

To get the number of diagonals of a seven-sided figure we multiply 7 by 4 and then divide by 2. It is characteristic of most problems in permutations and combinations that they involve multiplication, and often division as well.

To see why multiplication occurs so much, let us consider a particular question. The Morse code consists of dots, which are short squeaks, and dashes, which are long squeaks. When Morse was inventing his code, he must have considered whether he could cover the alphabet without his signals becoming too long. So he would arrive at the questions – how many signals can you make with two squeaks? – with three squeaks ? – with four squeaks? It is evident that, with one squeak, you can only make two signals, namely . and —.

When you are sending a series of dots and dashes, at each stage you have a choice; your next squeak may be a dot or a dash. We can show this pictorially as a dividing of the ways. Choosing a dot means taking the road to the left, a dash the road to the right. Figure 80 shows the Morse signals arranged in this way.

Now this figure is very much like the trees we used on pages 98 and 99 of *Vision* as a way of picturing multiplication. At each

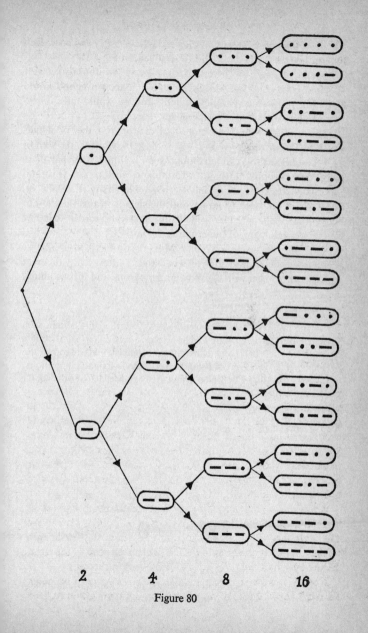

2 4 8 16

Figure 80

stage of the process, every branch grows two twigs, and so at each stage the number of end results is doubled. In Figure 80 we left off at the stage where there were sixteen signals with four squeaks. If we had gone on to five squeaks, each of our sixteen signals would bud out to produce two new signals, and we would have thirty-two signals at the right of the page.

Incidentally, in passing, note that the total number of signals with four squeaks *or less* is $2 + 4 + 8 + 16$, which is the sum of a geometrical progression, though such a simple one that there would be no point in using the formula to work it out. This total in fact is 30; as there are only twenty-six letters in the alphabet, Morse did not need to go beyond the signals in this diagram to cover the alphabet. He needed longer ones in order to cover numerals and signs of punctuation as well.

You may have noticed that Morse's problem in 1837 was essentially the same as Pingala's in 200 B.C. – how many different effects can you get with a given number of long and short sounds?

Exercises

1. In the Braille system of books for the blind, there are certain standard positions at each of which the paper may either be raised into a knob, or left flat. How many different signals could you convey to a blind person by means of 1 standard position? – 2 standard positions? – 3, 4, 5, 6 positions? (Braille in fact used 6, placed like the dots of the 6 used in dice.)

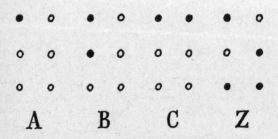

Figure 81

Braille type. The black circles indicate raised knobs. The hollow circles show where the paper is left flat.

2. In a football game, the result may be a win (W), a draw (D), or a loss (L) for the home team. A person predicting the results of four matches might write, say, WWDL. Draw a tree, resembling that for the Morse code, showing the possible predictions for 1 game, for 2 games, for 3 games. What rule gives the number of possible outcomes for *n* games? How many lists would you have to make, if you wanted to be sure that one of them predicted correctly the result for each of 7 games? If you knew nothing about football, what is the chance that a single entry would give the results of 7 games correctly?

In the Morse code, you are free to use as many dots and dashes as you like. But there are problems in which objects can be used only once. For example, if you have 4 flags, of different colours, how many different signals can you send by using 3 of them? You may fly the flags, say, one above the other, so the person reading the signal sees the flags in a definite order. Red, blue, yellow would convey a different signal from yellow, red, blue.

Figure 82

Let us denote our flags by A, B, C, D. We have only one flag of each colour, so a signal such as AAAA or DACC is not possible. But we still have a series of decisions to make, and we could represent them by branching roads. We have, of course, four choices of our first flag, it may be A, B, C, or D. But later choices will depend on what has been done earlier; if we have, for instance, used A for our first flag, we have only B, C, and D left in our hands to choose for the second. Actually, there are enough choices here to make the tree diagram rather cumbrous. Figure 82 is equivalent to the tree but takes rather less space and is (I believe) easier to follow. The large letters represent the flag first chosen, the medium-sized letters the flag chosen second, and the small letters the flag chosen third. If A has been chosen for the first flag, the second flag must be chosen from B, C, and D. These three possibilities are seen in the diagram, next to the big A. If we have chosen A first and B second, our third flag must be either C or D. These two letters will be seen following the big A and the medium B. In the same way, all the other possible choices are shown. Every three-flag signal can be found in the diagram by selecting a large letter, a medium letter placed to the right of it, and a small letter placed to the right of that medium letter.

You can, if you like, imagine the branches of the tree superimposed on this diagram. From the starting point, four branches would go out to the four big letters. From each of these, three branches would go out to the medium letters. From each medium letter, two branches would go out to small letters.

There are four possibilities for our first flag. When the next flag goes up, the number of possible signals has increased to 4×3. You can see these twelve signals in the diagram, if you imagine the small letters erased. With the third flag, each of the earlier signals branches out into two and we have in all $4 \times 3 \times 2$ possible signals.

We could of course work out $4 \times 3 \times 2$ and find it to be 24. This however is not particularly helpful. We are trying to find a pattern in our answers which will enable us to answer such questions as 'How many 8-flag signals could you make with 30 flags?' It does not help us at all to know that there are 24 signals that can be made with 3 flags chosen from a set of 4 flags. But it does help us a little to know that, with 4 flags, the number of 1-flag signals is 4, the number of 2-flag signals is 4×3 and the number of 3-flag signals is $4 \times 3 \times 2$. Here a very definite pattern is beginning to emerge. In the exercises below, the actual multiplying out of the answers is unimportant and indeed unnecessary. All the answers should be left in a form resembling the answer $4 \times 3 \times 2$ above.

Exercises

1. It would be tedious to draw the whole of the diagram corresponding to Figure 82 to show the signals that could be made with 3 flags

chosen from 5 flags A, B, C, D, E. However by drawing part of it one can see how to answer the following questions:

    (i) Figure 82 contains 4 distinct blocks. How many blocks would there be in the present case?

    (ii) How many medium-size letters occur in each block?

    (iii) How many small letters stand near each medium-sized letter?

    (iv) What expression replaces the $4 \times 3 \times 2$ which we found as the answer to our earlier question?

2. Do as much or as little experimenting as you find necessary to answer the following questions:

    (i) How many 3-flag signals could you make with 6 flags A, B, C, D, E, F? (ii) With 7 flags? (iii) With n flags?

3. How many 4-flag signals could you make with

    (i) 6 flags; (ii) 7 flags; (iii) n flags?

THE ROLE OF DIVISION

Questions 3 and 6 at the end of the chapter 'A Method of Discovery' were already preparing for the effect of division in permutations and combinations. On a railway with 4 stations the tickets needed are 12 in number. In a tournament with 4 players, each playing each, the number of games is 6, exactly half of 12. The reason, as we saw, was that A to B and B to A are two different tickets, but, in the game problem A playing B is the same as B playing A. Thus to each entry in a list of games there correspond two entries in a list of tickets. The list of games could be set out like this:

AB	AC	AD	BC	BD	CD

The list of tickets could be arranged thus—

AB	AC	AD	BC	BD	CD
BA	CA	DA	CB	DB	DC

Each entry in the second row is formed by reversing the order of letters in the entry above it in the first row.

    In this particular instance, one might naturally think the easiest procedure would be to list the six games first and then, if you wanted to know the number of tickets, to obtain it by doubling 6. If however it happened that one already knew the number of tickets to be 12, it would be equally possible to work in the opposite direction, and find the number of games by dividing 12 by 2.

It may appear unnatural but in fact this latter procedure is the one that will be most frequently used. The number of tickets on the railway line is the same as the number of two-flag signals possible with four flags, and we have already studied this question. In your work on the questions at the end of the last section you probably discovered the general pattern in the answers to all questions of this kind, *in which the order of the letters mattered*. It is usually not easy to find a systematic way of counting the possibilities in problems where order is unimportant. It has been found by experience that a good way to attack such problems is first to count the possibilities in a question where order *does* matter and then see in what way the possibilities merge together if order is ignored.

For example, a question where order does not matter is – in how many ways can you choose three children to go on a picnic from a group of six children? Here each child is simply chosen to go or to stay behind. But if the picnic were to be organized on authoritarian lines, with a captain, first officer, and a crew (of one), order would become important. How many ways would there be then?

Now we know the answer to the second question. It is the same as the number of three-flag signals with six flags. The number is $6 \times 5 \times 4$. This number is 120, and it would require rather a large amount of space to show in detail the way in which these 120 entries would fuse into groups if we changed to an equalitarian system and parties of three children went to the picnic simply as good friends. However we can give an indication of the effect. The 120 possibilities could be listed as follows:

ABC	ACB	BAC	BCA	CAB	CBA
ABD	ADB	BAD	BDA	DAB	DBA
....	....	....	....	....	....
DEF	DFE	EDF	EFD	FDE	FED

These are the 120 cases with rank counting. ABC means A is captain, B first officer, C crew, and is thus different from CBA, which represents a complete reversal of social precedence. In the complete list there would be twenty rows with six entries in each row.

Now in each row, the children are the same; it is only the ranks that differ. It we abolish rank and ignore order, it becomes exactly the same to say that C, B, and A may go the picnic as to say that A, B, and C may go. In each line, the six lists merge into a single list. The effect accordingly is to divide the number of possibilities by 6; instead of 120 lists, 120/6 will cover all possibilities.

What is the significance of this number 6 that we divide by? One might make a rash guess and think that we divide by 6 because 6 is the number of children in the group from which we choose. A little thought or experiment will show that this is not so. If we had only four children, A, B, C, D to choose from, we would still find we were dividing by 6. In fact the top two rows would look exactly as they do now.

No, the significance of 6 here is that it is the number of ways in which rank can be assigned to the same group of three children. Now from three children, we can choose a captain in three ways. Having chosen the captain, there are two children available for first officer. When captain and first officer have been chosen, there is only one candidate left for crew. Thus the number of ways of assigning rank is $3 \times 2 \times 1$, corresponding to the diagram of Figure 83.

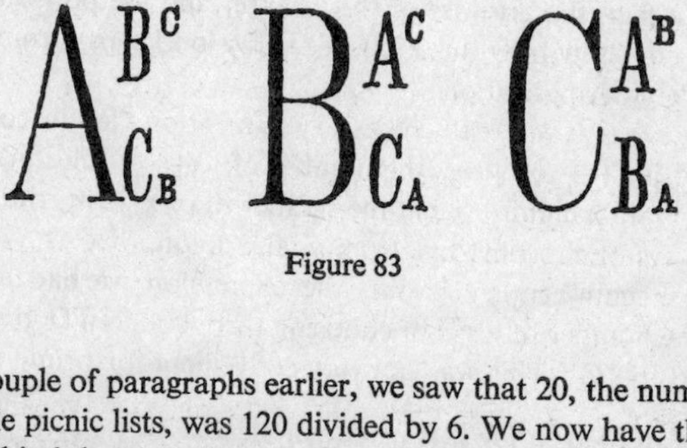

$$A^{B^C}_{C_B} \quad B^{A^C}_{C_A} \quad C^{A^B}_{B_A}$$

Figure 83

A couple of paragraphs earlier, we saw that 20, the number of possible picnic lists, was 120 divided by 6. We now have the pattern behind these numbers; our answer, 20, is $6 \times 5 \times 4$ divided by $3 \times 2 \times 1$. This is the number of ways of choosing three children from six. It will be noticed that both $6 \times 5 \times 4$ and $3 \times 2 \times 1$ contain three factors, corresponding to the number of children to be chosen. In both $6 \times 5 \times 4$ and $3 \times 2 \times 1$ we see numbers arranged in decreasing order. And in $6 \times 5 \times 4$, the first number is 6, the number of children from which we choose.

By observing these facts, it is easy to guess that if three children had to be chosen from ten, the number of possible lists would be $10 \times 9 \times 8$ divided by $3 \times 2 \times 1$. If three children had to be chosen from n children, this could be done in $n(n-1)(n-2)$ divided by $3 \times 2 \times 1$ ways. It is good to make this guess, but it is also good to see that it can be justified by exactly the same argument that we used for the particular case of three children chosen from six. The best way to become familiar with this type of result is repeatedly to think through the procedure and the argument that brings us to it.

As $3 \times 2 \times 1$ is the same as $1 \times 2 \times 3$, the number of ways of choosing three children from n may be written

$$\frac{n(n-1)(n-2)}{1.2.3}$$

This expression is not new to us. Expressions of this kind (with x instead of n) provided the climax in the chapter 'A Method of Discovery'. Again, at the beginning of this chapter, in the Binomial Theorem the coefficient of x^3 was precisely the expression above. In fact, the purpose of this chapter and the previous one has been to show how this kind of expression keeps turning up, in a variety of applications.

I hope it is now not really necessary to mention that the coefficient of x^4 in $(1+x)^n$ gives the number of ways of selecting four children from n children, that the coefficient of x^5 gives the number of ways of choosing five from n, and so on.

In nineteenth-century books, the expression we had above, giving the number of ways of choosing three children from n, was indicated by $^nC_3$. This was not very convenient for printers, and in books printed today it is usually indicated by $\binom{n}{3}$.

Thus, for a whole number n the Binomial Theorem may be found in the form

$$(1 + x)^n = 1 + \binom{n}{1}x + \binom{n}{2}x^2 + \binom{n}{3}x^3 + \ldots + \binom{n}{n}x^n.$$

Now on the right-hand side of this equation it is noticeable that every term except the first contains one of these new bracket symbols. This makes us think that perhaps we could bring the first term into line with the others. The pattern of the equation suggests that the first term should be $\binom{n}{0}$, which would mean the number of ways of choosing a picnic party of none. There is only one way to do this – keep all the children at home. Accordingly $\binom{n}{0}$ is interpreted as 1, and, with this understanding, all the co-efficients of $(1 + x)^n$ can be expressed by these bracket symbols.

A PROOF WITHOUT CALCULATIONS

On page 200 we arranged the Binomial Coefficients in the Pascal (or Pingala) Triangle. Two rows of this triangle, corresponding to $(1 + x)^4$ and $(1 + x)^5$ were

$$\begin{array}{ccccccc} & 1 & & 4 & & 6 & & 4 & & 1 \\ 1 & & 5 & & 10 & & 10 & & 5 & & 1 \end{array}$$

If we wanted to refer to these numbers by means of the bracket notation just explained, they would appear as

$$\binom{4}{0} \quad \binom{4}{1} \quad \binom{4}{2} \quad \binom{4}{3} \quad \binom{4}{4}$$

$$\binom{5}{0} \quad \binom{5}{1} \quad \binom{5}{2} \quad \binom{5}{3} \quad \binom{5}{4} \quad \binom{5}{5}$$

Now the numbers in this triangle, as was mentioned on page 200, can be calculated by making each number the sum of the two numbers immediately above it. Thus, in the row 1, 5, 10, 10, 5, 1 the second 10 is the sum of the numbers 6 and 4 that stand above it. In terms of our bracket symbols this means $\binom{5}{3} = \binom{4}{2} + \binom{4}{3}$

Since every number in the triangle is the sum of two numbers standing above it, this last equation is a particular instance of a general rule. If we replace 4 by any number n (so that 5 is replaced by $n + 1$) and 2 by any number r (so that 3 is replaced by $r + 1$) we get the general result $\binom{n+1}{r+1} = \binom{n}{r} + \binom{n}{r+1}$

This equation expresses the procedure by which the numbers in the triangle are computed. Now presumably we ought to prove that this procedure is justified. This can be done purely by the routines of algebra. We can replace the bracket symbols by the formulas for which they stand, and verify that the equation is true. This is not really hard, but neither is it exceptionally easy; one does meet people who cannot do it. J. L. Thomas, who taught at Highgate School, had a colourful way of obtaining this result without any calculation at all. He approached it by means of the traditional school story. A school has $n + 1$ members; one boy, who is the hero, and a mob of n others. The story, of course, is concerned with whether the hero will get into the school team or not. There are $r + 1$ places in the team. As these $r + 1$ places have to be filled from the $n + 1$ boys in the school, there are $\binom{n+1}{r+1}$

ways of selecting the team. (This is the left-hand side of the equation we hope to prove.) Now either the hero is in the team or he is not, so the list of possible teams can be broken into two parts – the cases where the hero is in the team, and the cases where he is out. Suppose he is in. Then there are r places still to be filled from the other n members of the school, and this can be done in $\binom{n}{r}$ ways. This is the number of cases where he is in. On the other hand, suppose he is out. Then the whole team, $r + 1$ places, has to be filled from the other n boys. This can be done in $\binom{n}{r+1}$ ways, which accordingly is the number of cases where the hero is out. If we now write the equation stating that the total number of possibilities is got by adding the number of cases where the hero is in to the number where he is out, we obtain the equation we want to prove.

Discovering the Binomial Theorem

Some years ago I knew a mathematician who had a horror of long calculations. His favourite dictum was, 'When you are calculating, you do not know what you are doing. When you really understand something, you say – "obviously this is so".' The proof just given may perhaps serve as a very elementary example of what he meant.

CHAPTER TEN

Algebra as a Key to Geometry

THE exact title of this chapter should be noted. It presents algebra not as *the* key to geometry, but as *a* key. There are many ways of handling geometry. Each time you learn a new way, you become enthusiastic about how well it handles certain problems and you begin to think perhaps this is the magic method before which all problems instantly surrender. After a little while you realize this is not so; the method deals very well with a certain class of problems; for other problems, other methods are more effective. It is wise therefore to have some idea of as many different approaches as possible.

The possible approaches fall into two main families. One family, named the synthetic approach, goes back to Euclid. In it, we draw pictures, state certain assumptions, and reason from these. In recent times, some geometers have not even drawn pictures. The essential feature of this approach is that it uses arguments but not to any extent calculations. The second approach is the analytic or algebraic. It goes back to Descartes, makes use in one way or another of the idea of graph paper, and proceeds by writing equations and making calculations.

These two methods complement each other. The advantage of the approach through diagrams and argument is that you see what you are doing. The pictures suggest results; it seems 'obvious' that this line is as long as that one, or that a certain angle is a right angle. Sometimes the pictures even suggest how to prove the result you suspect to be true. But sometimes they do not, and this is the great weakness of the synthetic approach. It is entirely unsystematic. Anyone who studied geometry at school by methods related to Euclid's will have had the experience of looking at some result to be proved and not having the faintest idea of how to begin. And indeed one could work at a problem of this kind for years without any certainty of ever finding a proof. This seems to have been the kind of consideration that led Descartes to look for another way of handling geometry, the algebraic. In that approach, if you want to prove that two lengths or two angles have the same size, you do not wait for an inspiration. You begin calculating the sizes in question. In doing this, you have a considerable amount of routine processes at your disposal. The calculations may turn out to be long and involved, so that someone who does not have an electronic computer specially programmed to carry out algebraic operations may give up from fatigue, or from a realization that he will be an old man or even die before he gets to the end of the computation. But algebra does have this steamroller quality; in principle the calculation can be carried through to the end; it is not necessary to wait for an inspiration.

The weaknesses of the algebraic approach are twofold. It can sometimes happen that the algebraic calculation is extremely long and heavy, while the result can be proved very simply and easily from the diagram. It can also happen that errors are made in calculation. For in a long calculation you can rarely see what you are doing. You are following certain procedures and, as your brain becomes tired, you are very liable to make mistakes; *plus* is written where it should be *minus* or you make some insane mistake in elementary arithmetic. This is quite sufficient to make the final answer completely misleading. If, by reasoning from pictures, I had come to feel strongly that a certain thing was true, and calculations then seemed to show that it was not true, my first thought would be to look for an error in my calculations. But of course the error might be in the reasoning. Mistakes in logic are usually due to overlooking some possibility, and such a mistake in thought could be revealed by a calculation. I might have thought something could happen in only one way; the calculation might lead to an equation with several solutions, one of which agreed with my original picture of the situation. In this case, it might well be that the calculations were right. I would look back at the diagram and see if the new solutions made sense geometrically. In short, one should trust neither argument nor calculation implicitly, but continue checking one against the other until they both give the same conclusion.

The Search for Pattern

THE BRIDGE FROM GEOMETRY TO ALGEBRA

Vision, the first book in this series, was mainly about algebra. However, care was taken to include in it a number of ideas and results which, taken together, allow any ordinary problem in geometry to be translated entirely into a question in algebra. In Chapter 11 of *Vision* we saw that the position of a point could be specified by giving two numbers, so we could speak of the point (3, 4) or quite generally of the point (x, y). In the same chapter we met the idea that an equation, such as $xy = 12$, could paint a picture for us on graph paper. We supposed that all points such as (12, 1), (6, 2), (3, 4), (−2, −6), (5, 2·4), which are examples of (x, y) such that $xy = 12$, receive a spot of black paint. All other points receive a spot of yellow paint. The points that satisfy the equation thus appear in black against a yellow background. In this case the black points form a curve which falls into two parts. The equation $xy = 12$ is called *the equation of the curve* or *the condition that a point lies on the curve*.

In geometry we are very often concerned with the point where two curves meet. In an earlier section of this book, 'On the Abundance of Solutions' (page 102), we showed how this situation was reflected in algebra. The point (x, y) where the curves cross must satisfy the equations of both curves. This idea is perpetually used in the algebraic treatment of geometry; one needs to meditate on it until it becomes second nature.

A basic idea in Euclidean geometry is that of *distance*. (The word *Euclidean* is needed here, because there are geometries, such as projective geometry and topology, in which distance plays no role.) Suppose some part of the earth's surface is flat and is small enough to be regarded as a plane. If we were told the names of the towns and villages in this region, and the distance (as the crow flies) from each to each, we could draw an accurate map of the region. All the information needed for the map is contained in the table of distances. It might happen that a group of villages formed some kind of geometrical figure. The precise shape of that figure, and all its geometrical properties, could be determined by examining the table of distances.

Everything in Euclid's geometry, then, can be expressed in

terms of distance. If we have a formula for finding distances, we can translate everything in Euclid's geometry into algebra. This is one of the reasons why it was important to have in *Vision* a section providing such a formula, namely Chapter 6, which dealt entirely with Pythagoras' Theorem.

How does this theorem enable us to find a formula for distance? We consider a particular case first. In Figure 84, A is the point (2, 1) and B is (9, 5). How long is the line AB? It will occur fairly readily to most people that in that figure ACB is a right-angled triangle, and we can use Pythagoras to find AB. For we have $AC = 7$ and $BC = 4$, so $AB^2 = AC^2 + BC^2 = 49 + 16 = 65$. Thus $AB = \sqrt{65}$, a little more than 8.

Particular cases such as this are easy to deal with. The method can be fairly primitive; we can, and many no doubt in fact will, find the distance AC by counting along the divisions between A and C. This procedure would become very inconvenient if the numbers were large, and quite impossible if the positions of A and B were specified by algebraic symbols. We therefore need to ask how the result $AC = 7$ is related to, and by what mathematical procedure it could be derived from, the information that A is (2, 1) and B is (9, 5).

We can get from O to any point in the plane by going a certain distance east and then a certain distance north (minus numbers being acceptable, if needed). For A, the distance east is 2; for B, it is 9. It should become clear, after some consideration of Figure 84, that 7, the length of AC arises as the difference of these numbers. Similarly, the distances north are 1 for A and 5 for B (as shown at the side of the figure), and the length BC is the difference between these. In fact we may set the situation out like this:

	East	North
B	9	5
A	2	1
Differences	7	4

From the numbers so found we then form $7^2 + 4^2$. which gives AB^2.

Our next job, after extracting this procedure, is to find a formula that will express it briefly and conveniently, for any pair

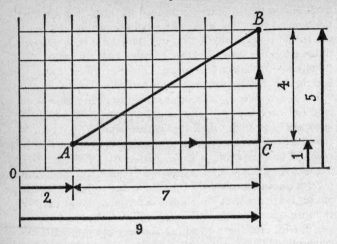

Figure 84

of points A and B. We could, if we liked, suppose the points to be (p, q) and (r, s). This would have the disadvantage that p, q, r, s do not suggest the coordinates of A and B, and in using the formula we might well get mixed up as to which number each of these letters stood for. We have strong associations between x, y and coordinates, but here we run into a difficulty; if we use (x, y) for A, what are we to use for B? We can get round this by using subscripts, taking x_1, y_1 as the coordinates of A, the first point, and x_2, y_2 as the coordinates of B, the second point. Replacing the particular numbers $(2, 1)$ and $(9, 5)$ by these symbols we arrive at the scheme

	East	North
B	x_2	y_2
A	x_1	y_1
Differences	$x_2 - x_1$	$y_2 - y_1$

The differences correspond to the numbers 7 and 4 we had earlier, and to $7^2 + 4^2$ there now corresponds $(x_2 - x_1)^2 + (y_2 - y_1)^2$, which accordingly gives AB^2.

This result will be referred to as the *distance formula*. It will be

230

used repeatedly, and is so well known that often it is not even mentioned. If you are reading a calculation in a book or journal, you should not expect to find, 'A is (2, 1) and B is (9, 5) so by the distance formula AB = $\sqrt{65}$' but rather 'A is (2, 1) and B is (9, 5), so AB = $\sqrt{65}$'. The reader is expected to know how to find the distance between two specified points.

The best way to memorize this formula, or procedure, is by repeatedly working out particular examples and re-thinking the argument by which the general formula is obtained.

For the points A and B in Figure 84, the formula gives AB^2 = $(9 - 2)^2 + (5 - 1)^2$. In each bracket we find a positive number: $9 - 2$ is $+7$ and $5 - 1$ is $+4$. Now of course, we might have taken the points in a different order, with (9, 5) as the first point and (2, 1) as the second. In this case our formula would have led us to the expression $(2 - 9)^2 + (1 - 5)^2$, which gives $(-7)^2 + (-4)^2$. Here again we see how conveniently things work out as the result of our agreement that 'minus times minus is plus'. The minus signs do not cause any trouble, since $(-7)^2 = 49$ and $(-4)^2 = 16$. We still get exactly the same answer, $49 + 16 = 65$, for the square of the distance. Something would be seriously wrong if we did not; our formula would then imply that the distance from A to B was different from the distance from B to A.

The numbers that occur inside the brackets, $x_2 - x_1$ and $y_2 - y_1$, are positive only when the second point lies both to the east and to the north of the first point. However in all cases the same formula gives the distance correctly. If you feel doubtful about this assertion, it is good to experiment with some of the other cases. For example you could draw the diagram with the first point (4, 1) and the second point (1, 5) and compare the distance found by reasoning from the diagram and the distance found from the formula. It is essential, of course, to remember that the square of a minus quantity is positive.

There is relatively little new in what has just been said. Already on page 138 of *Vision* a method was given by which the distance between any two points on graph paper could be found, and this method was used in the exercises a few pages later. Only two new ideas have now been added. First, instead of applying the method only to particular points, we have applied it to (x_1, y_1) and (x_2, y_2), and so have obtained an explicit formula instead of a procedure to be followed. Secondly, we have observed that this one formula applies to all cases; the minus signs which may appear do no harm.

CIRCLES

On page 281 of *Vision* we met an equation whose graph was a circle. Earlier in this book, in the section 'On the Abundance of Solutions', some mention was made of the equation of a circle. However, in neither of these places did we prove what the equation of a circle had to be. Now we are in a position to do this, and it is not difficult; it follows directly from the distance formula.

How is a circle defined? What do we mean by a circle? Circles are often drawn by using compasses, but this rather obscures the central idea. A more primitive method gives a better clue. Suppose you want to make a circular flower bed in a garden or park. You drive a stake in, fasten a length of rope to it and walk round the stake. As you go, the free end of the rope traces out the border of your flower bed. All the points on the edge of the flower bed are thus the same distance from the stake in the centre, and this is the defining property of a circle. If, for example, we want the circle with centre (3, 4) and radius 5, we must mark every point that is at a distance 5 from the point (3, 4). If we want to know whether some point (x, y) is on the circle, we have to ask the question, 'Is the distance of (x, y) from (3, 4) equal to 5?' Now the distance formula tells us that the square of the distance between (3, 4) and (x, y) is $(x - 3)^2 + (y - 4)^2$. The point (x, y) will be on the circle if this is the same as the square of 5, namely 25. So the point will be on the circle if $(x - 3)^2 + (y - 4)^2 = 25$. This then is the equation of the circle. Depending on what you are doing, you may wish to leave the equation in this form, or you may prefer to multiply it out. Multiplied out, the equation becomes $x^2 + y^2 - 6x - 8y = 0$. You may observe that this equation is satisfied by (0, 0), so the origin, O, apparently is on the circle. Is this right? Should it be?

You may notice that, in making up this last example, I made use of the 'Egyptian' right-angled triangle with sides 3, 4, 5. This means that there are many points on the circle with simple co-ordinates. If you start from the centre and go 4 east and 3 north, you will get a point on the circle. What are the coordinates of this point? Do they satisfy the equation? Similarly you could go from the centre 4 east and 3 south, or 3 west and 4 north, or in several other ways make sure of getting a point at distance 5 from the centre (3, 4). It is perhaps a useful exercise to mark all these points on graph paper, record their coordinates, and check that they do satisfy the equation of the circle, and both the forms given above.

One feature of the exercise just given is that you know, on geometrical grounds, how it ought to turn out. If you make a slip in algebra or arithmetic, you will detect it. This is a rather simple illustration of the remark made at the beginning of this chapter, that accuracy of thought and accuracy of calculation can be built up by continually checking the algebraic computation against the geometric picture.

Exercises

1. (i) What is the square of the distance of the point (x, y) from the point $(1, 1)$? (ii) What equation must (x, y) satisfy if its distance from $(1, 1)$ equals 1? Give the answer in both forms, i.e. as it first arises, and also multiplied out. (iii) Sketch the circle with centre $(1, 1)$ and radius 1. What four points, with whole number coordinates, lie on this circle? (iv) Do the four points found in (iii) satisfy the equation found in (ii)? Should they?

2. (i) What is AP^2 if A is $(2, 1)$ and P is (x, y)? (ii) What is the equation of the circle with centre A and radius 3? Give both forms. (iii) Plot the following points: $(5, 4)$, $(5, 1)$, $(5, -2)$, $(2, 4)$, $(2, -2)$, $(-1, 4)$, $(-1, 1)$, $(-1, -2)$. Which of them lie on the circle specified in (ii)? (iv) What would be the radius of a circle, centre A, drawn to pass through the remaining points (i.e. those not mentioned in your answer to (iii))?

3. Find the equation of the circle with centre $(5, 5)$ and radius 5. Draw this circle on graph paper with the help of compasses. Through what points with whole number coordinates does it pass?

4. (i) What is the square of the distance of the point (x, y) from the point (a, b)? (ii) What equation must (x, y) satisfy if its distance from (a, b) equals r? (iii) Of what curve is the answer to (ii) the equation?

5. (i) What is the square of the distance from (x, y) to the origin $(0, 0)$? (ii) What is the equation of the circle, centre $(0, 0)$, radius r?

PERPENDICULAR LINES

In Figure 85, A is the point (2, 1) and O is the origin (0, 0). Suppose for some reason we want to find the line through O perpendicular to OA. To do this, we need only to know Pythagoras' Theorem and the distance formula. Suppose B is the point (x, y).

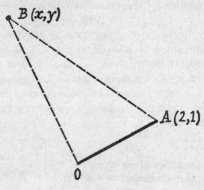

Figure 85

We want to find what equation (x, y) must satisfy in order to make OB perpendicular to OA. If OA and OB are perpendicular, the triangle OAB is right-angled, and by Pythagoras we shall have $AB^2 = OA^2 + OB^2$. Accordingly, all we have to do is to use the distance formula, to give us AB^2, OA^2, and OB^2 in terms of the coordinates involved. When we have done this the equation becomes

$$(x - 2)^2 + (y - 1)^2 = (2^2 + 1^2) + (x^2 + y^2)$$

If you multiply this out, you will find it simplifies considerably, and leads to the result $2x + y = 0$. This, as we expected, is the equation of a straight line.

A standard result, given in most textbooks, is the rule for finding the line through O perpendicular to a given line, $y = mx$. Instead of reading the section of the textbook that derives the rule, it is good to pose the question, 'How would I find the line through O perpendicular to $y = mx$?' The method of the previous paragraph is perfectly adequate. If we take A to be the point $(1, m)$ then OA will be the line $y = mx$. If, with this new A, we now proceed to find the equation of the line OB perpendicular to OA, our problem is solved. You may care to do this before reading on.

*   *   *

It often happens that, after finding the solution of a problem by one method, we see by examining the solution that it could have been found by some other method. This happens with the two questions just discussed. If we draw the line $2x + y = 0$, found in the first question, we find that it passes through the point $(-1, 2)$, which is called P in Figure 86. Two triangles are shaded

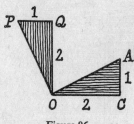

Figure 86

in the figure; they look rather like two sails of a windmill. It is clear that the triangle OQP could be obtained by swinging the triangle OCA through 90°, for OC must turn through 90° to reach the position OQ. Accordingly OA must swing through 90° to reach the position OP. This checks then that we have found the correct line. It would be possible, by starting out from these considerations, to find the perpendicular line without any mention of Pythagoras or the distance formula. We will now proceed to do this with the other question.

In Figure 87, we begin with the point A, $(1, m)$, and shade the triangle OCA which emphasizes that A is reached by going 1 to the east and m to the north from O. We then suppose this shaded triangle swung bodily through 90° to reach the position OQP.

Accordingly OP is the line perpendicular to OA. To get from the origin to P we evidently go 1 to the north and then m to the *west*. P is thus the point $(-m, 1)$. A moment's reflection may be needed to see this, since in going from O to P via Q, we were slightly unconventional; we made the journey in the y-direction first and in the x-direction second. The line OP is $x + my = 0$, or $y = -\left(\dfrac{1}{m}\right)x$.

In Figure 85, the line OA rises half a unit for every unit we move across and is accordingly said (by various writers) to have steepness, slope or gradient $\frac{1}{2}$. On the line OP, if we move 1 unit to the right, we *sink* by 2 units, and the steepness of OP is thus -2. It will be noticed that there is a connexion between the two numbers occurring here, $\frac{1}{2}$ and -2. Multiplied together they give -1. That this is not an accident can be seen from Figure 87. The steepness

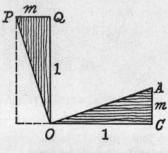

Figure 87

of OA is m, of OP is $-1/m$, since it falls by 1 unit as we go m units across. Multiplying together m and $-1/m$ we do get -1, and since m here can be any number whatever we see that we have a perfectly general rule.

In textbooks this result often appears in the form that $y = mx$ and $y = m'x$ are perpendicular if $mm' = -1$. It may also appear as $m' = -1/m$.

On pages 308 and 309 of *Vision* an activity was described which gives a feeling for the relation between the number m in an equation such as $y = mx + c$ and the steepness of the corresponding line.

THE REGULAR HEXAGON

At the beginning of geometry learners are often shown how to draw a regular hexagon. This depends on the perhaps surprising fact that, if you draw a circle and then – with great care – repeatedly step off a distance equal to the radius, this distance fits exactly six times into the circle, as shown in Figure 88(a). Children often elaborate on this construction to obtain the six-petalled flower shown in Figure 88(b).

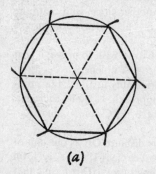

(a) *(b)*

Figure 88

Synthetic geometry provides a perfectly satisfactory explanation. One proceeds first by showing that the angles of a triangle give a total of 180°. In an equilateral triangle, all the angles are equal, so each angle must be 60°. Accordingly, six such angles will make 360°, and six equilateral triangles will fit together to fill all the space near the centre of the circle, and form a regular hexagon. Even so, it is of interest to examine this question from the algebraic viewpoint. The algebraic investigation will give us some results which do not appear in the synthetic argument, and will also lead us in a natural way to consider a more general question.

Figure 89 shows a convenient way of constructing the regular hexagon. DOA is a straight line through O, the centre of the circle.

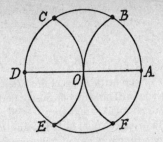

Figure 89

Part of a circle is then drawn with centre A, to pass through O and cut the original circle in B and F. Another circle, with centre D, similarly gives us the points C and E. This construction ensures that each of the lines AB, AF, DC, DE will have the same length as OA. What is not obvious, and what we wish to prove, is that the lines BC and EF also are equal to OA in length.

We do not need any subtlety or originality to attack this problem by algebra. A sledgehammer approach is quite sufficient. We can work out the coordinates of B, F, C, and E, and from these find the lengths BC and EF.

It will be convenient to choose the distance OA as our unit of length. Thus the radius of the original circle is 1, and A is (1, 0). The point B is found where two circles cross. This means that OB = 1 and AB = 1, so we have two equations concerning B, which (unless some exceptional situation is involved) should be enough to determine two unknowns, the coordinates of B. Since $OB^2 = 1$ and $AB^2 = 1$, the distance formula allows us to express these equations in terms of (x, y), the coordinates of a point where the circles cross. This last phrase had to be chosen with a certain care. I did not want to say that (x, y) must be the co-ordinates of B, since the circles cross also at F, and it is just as true of F as it is of B that the distance from O is 1, as also is the distance from A. The equations we obtain will in fact give two solutions, one of which will correspond to B, the other to F.

Accordingly, if (x, y) stands either for the point B or the point F, the distance formula tells us that x and y satisfy both the following equations: $x^2 + y^2 = 1$ and $(x - 1)^2 + y^2 = 1$. Now we have not had any special method for solving such a pair of equations; fortunately, no special method is needed. It is natural to multiply out the second equation, and as soon as this is done something leaps to the eye. The second equation gives $x^2 - 2x + 1 + y^2 = 1$. As the first equation is $x^2 + y^2 = 1$, everything in the first equation occurs in the second equation, together with $-2x + 1$ on the left-hand side. Accordingly if we subtract the first equation from the second, we shall be left with $-2x + 1 = 0$, which means $x = \frac{1}{2}$. This result looks reasonable on geometrical grounds. The point B looks as if it lies directly above F, and joining BF we get a line that looks as if it bisects OA. This is as it should be if both B and F have the same x coordinate, namely $\frac{1}{2}$. If we now go back to our first equation, $x^2 + y^2 = 1$, and put $x = \frac{1}{2}$, we get the equation $y^2 = \frac{3}{4}$. As both $\frac{1}{2}\sqrt{3}$ and $-\frac{1}{2}\sqrt{3}$ when squared give $\frac{3}{4}$, y might be either of these numbers.* If we go back to our original equations, we can check that both

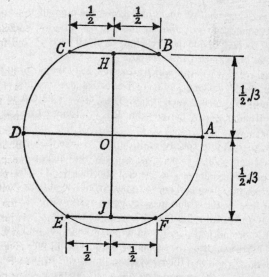

Figure 90

* For manipulation of square roots, see Chapter 14.

($\frac{1}{2}$, $\frac{1}{2}\sqrt{3}$) and ($\frac{1}{2}$, $-\frac{1}{2}\sqrt{3}$) satisfy these two equations. The first of these in fact represents B, the second F.

In exactly the same way, by considering the fact that each of the points C and E is distant 1 from 0 and also from D, we can show that C is ($-\frac{1}{2}$, $\frac{1}{2}\sqrt{3}$) and E is ($-\frac{1}{2}$, $-\frac{1}{2}\sqrt{3}$).

We could now use the distance formula to verify that the distances BC and EF are both 1, but this is rather using a steam hammer to crush a beetle. Figure 90 shows graphically what we have discovered about the coordinates of B, C, E, and F. It is evident from this figure that B and C are at the same level, and that the distance between them is 1. A similar remark applies to E and F. We have thus proved what we set out to do, that six points, A, B, C, D, E, F, can be spaced out around the circle of radius 1, in such a way that each point is at distance 1 from its neighbours.

A GENERAL PROBLEM

In the measurement of angle in degrees, the circumference of a circle is supposed to be divided into 360 equal parts. The points A, B, C, D, E, F divide the circle into six equal parts. Accordingly the angle BOA must be of size 60°.

There are two ways in which we can specify the position of a point in the plane. One way is to give its distance from the origin and the direction in which it lies. Thus, if we take OA as our standard direction, we can say that B is at distance 1 and at the angle 60°.

The other way to specify position is of course that associated with ordinary graph paper. We can say that B is reached by starting at O and going $\frac{1}{2}$ to the east and then $\frac{1}{2}\sqrt{3}$ to the north.

Now we very frequently need to pass from one of these systems to the other, just as a ship may set a course 30° north of east and travel 100 nautical miles (which corresponds to the first system) but wish to report its position in terms of longitude and latitude (which corresponds to the second system). It is not only in problems of navigation that the two systems interlock; in any problem dealing with the shapes and sizes of objects, we are likely at some stage to have information in terms of one system which we need to express in terms of the other system.

There is a fact which we can use to reduce the labour of solving this problem. Suppose a point P lies in the same direction as B (that is, at 60°) but at a distance 2 from the origin. It is not difficult to see that the coordinates of P will be exactly twice those of B. (Figure 91.) In the same way, if P were five times as far away as B, but still in the same direction, we could find the co-ordinates of P simply by multiplying each coordinate of B by 5. If P were *r* times as far away, we would multiply by *r*.

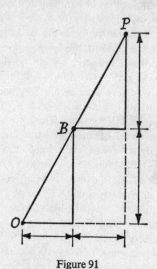

Figure 91

This means that we can deal with points at any distance if we can deal with points on our circle of radius 1. We have already succeeded in finding the coordinates of the special points A, B, C, ... at angles 0°, 60°, 120°, ... The general problem that now confronts us is *to find the coordinates of the point at any specified angle on the unit circle.*

Now of course it is not necessary for each individual to work out the answer to this question. This is a classical problem. Once the answer has been worked out, it can be preserved and made available in the form of tables, together with some theoretical formulas. This problem was in fact solved by Indian astronomers somewhere around 550 A.D. Improved methods became available later, after the discovery of the calculus, but that does not concern us at present. Essentially, this is a report on the approach of the Indian astronomers. It does not keep to the original form of their work, since the present symbolism of algebra and the use of coordinates came long after them. What it does bring out is something rarely emphasized in textbooks – that the whole problem is basically one of algebra, and that any difficulties that may arise in it are difficulties of handling and solving equations. This glimpse of history shows why the solving of equations was a central theme in mathematics for many centuries. Most of the equations are quadratics, and this goes a long way towards explaining the emphasis formerly put on quadratic equations. Knowing this history may help an individual to decide to what extent he needs to acquire skill in the handling of traditional algebra, for the work on this classical problem gives a fair sample of how algebra is applied to geometry. These methods are relevant for anyone involved in the design or construction of actual physical objects, which must have the right shape, fit together properly and perform some desired function. This work is relevant to any study which involves the movements of objects or the forces between them. It is also relevant to any subject which does not deal with the positions of actual objects, but which makes extensive use of diagrams and graphs. Probably the best way for anyone to decide how much of it he wants to learn is to look through a book which gives an adequate account of his subject, and see how much of this kind of algebra enters into it.

We now return to the consideration of our problem, and consider another simple question, that of finding the coordinates of the point at angle 45° on the unit circle. This is the point K in Figure 92. It is half way from A, the point $(1, 0)$ at 0°, to G, the point $(0, 1)$ at 90°. This means that the distance GK equals the distance AK. If K is (x, y) we can use the distance formula to convert the equation $GK^2 = KA^2$ into an equation involving x and y. We obtain in fact $x^2 + (y - 1)^2 = (x - 1)^2 + y^2$. When this is multiplied out, we find $x^2 + y^2 + 1$ occurs on both sides. Subtracting this amount from each side, we are left with $-2y = -2x$, which means that $y = x$. Since K is on the unit circle, $x^2 + y^2 = 1$. Substituting for y, using $y = x$, we find $2x^2 = 1$.

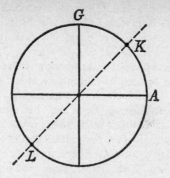

Figure 92

This means $x^2 = \frac{1}{2}$, so there are two possibilities, $x = 1/\sqrt{2}$ and $x = -1/\sqrt{2}$. Since K lies on the east side of O, it is clearly the first of these we want. (The second would lead us to L, at angle 225°, which is also equally distant from G and A.) As $y = x$, the value of y must also be $1/\sqrt{2}$. Thus K, the point at 45°, is $(1/\sqrt{2}, 1/\sqrt{2})$.

Now this last result is one that could be obtained fairly easily without using the machinery of coordinate geometry, so it may be felt that our method is unnecessarily heavy. However the procedure followed has a value not only as producing this particular result, but also as indicating what could be done in other cases. When we first translate $GK^2 = KA^2$ into algebra, we get an equation that is full of squares, and that accordingly looks as if it will not be too easy to handle. But as soon as we multiply out, a great simplification is seen, and we arrive at the very simple equation $y = x$. It is natural to wonder, is this because we are dealing with the two rather special points G and A, or is it something that would happen when we look for a point equally distant from *any* two points?

A beginner in algebra can find the answer for himself. He may experiment with particular points (which need not be on the circle). What is the condition that (x, y) shall be equally distant from $(1, 2)$ and $(3, 4)$? Does it simplify? After a few experiments of this kind, which will give a feeling for the procedure and strongly suggest a certain conclusion, the general problem may be tackled;

what is the condition that (x, y) shall be equally distant from (a, b) and (c, d)? Does it simplify?

*   *   *

Our question, then, can be answered by calculation. But the answer can also be seen by purely geometrical considerations. Suppose that in some emergency it was necessary to improvise a ruler with a straight edge, and all we had was a piece of torn paper with an irregular outline. What would we do? The natural solution would be to fold the paper. If a flat sheet of paper is carefully folded, the fold will always be a straight line. Now suppose we stick a pin through our folded paper, and open it out. The result will resemble Figure 93, in which S and T are the pinholes.

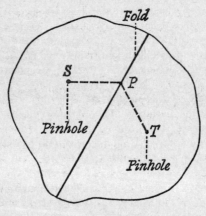

Figure 93

The dotted lines join S and T to any point P on the fold. When the paper was still folded, these dotted lines coincided, so naturally they have equal lengths. Accordingly, any point on the fold is equally distant from S and T. It seems reasonable that the points on the fold should be the only ones for which this is true; if for instance, P were on the same side of the fold as S, one would expect SP to be shorter than PT.

It is in fact a well-known theorem in Euclid's geometry, that the points which are equally distant from any two points, S and T, all lie on a line (which in fact is perpendicular to ST and bisects ST). Accordingly, it would be very surprising if our equation, expressing GK = AK, did not boil down to the equation of a straight line.

If S and T happened to be two points on the circle with centre O (the case that particularly interests us), the line would go through O, since each of OS and OT is a radius of a circle, so OS = OT. This means that O satisfies our condition, and so must lie on the line. Now any line through the origin has an equation of the form $y = mx$ and it is easy to find where this line meets the circle $x^2 + y^2 = 1$; we simply substitute mx for y in the equation of the circle. We get $x^2(1 + m^2) = 1$, and solving for x requires only the taking of a square root. (Anyone finding any difficulty here should adopt our usual recommended procedure; go back to particular cases: where does $y = 2x$ meet the circle? where does $y = 3x$ meet it?)

Accordingly, if we know the coordinates of two points on the unit circle, all we need to find the coordinates of the point on the circle midway between them is a table of square roots.

We can make up an example of this, that will only involve simple arithmetic, by using the 3, 4, 5 right-angled triangle. Dividing each of these numbers by 5, we see that a triangle with sides 3/5, 4/5, 1 will be right-angled. This means $(3/5)^2 + (4/5)^2 = 1$. Now $3/5 = 0.6$ and $4/5 = 0.8$. So if we take $x = 0.8$, $y = 0.6$ we have $x^2 + y^2 = 1$, and accordingly $(0.8, 0.6)$ is a point on the unit circle. Let us find the point on the circle midway between $(0.8, 0.6)$ and $(1, 0)$. If (x, y) is a point equidistant from the two points just mentioned we must have $(x - 0.8)^2 + (y - 0.6)^2 = (x - 1)^2 + y^2$. When we work this out, we find $x^2 + y^2 + 1$ occurs on both sides of the equation. Subtracting this amount, we are left with $-1.6x - 1.2y = -2x$. If we multiply both sides by -1, this becomes $1.6x + 1.2y = 2x$, from which it follows that $1.2y = 0.4x$. Dividing both sides by 0.4, we get $3y = x$. This is the equation of a line. It is in fact the equation of the line that would be obtained by folding the paper so that the point $(0.8, 0.6)$ was brought on top of the point $(1, 0)$. This equation completely expresses the fact that (x, y) is equidistant from our two points. To proceed any further, we must combine this with our other piece of information, that (x, y) lies on the unit circle, which means that $x^2 + y^2 = 1$. Clearly the thing to do is to substitute for x, using $x = 3y$. This means that $x^2 = 9y^2$, and gives us $10y^2 = 1$. So $y^2 = 0.1$. We want the positive value for y. A table of square roots gives $y = 0.3162$. Since $x = 3y$, the value of x must be 3 times this, namely 0.9486. Accordingly the point midway from $(1, 0)$ to $(0.8, 0.6)$ on the circle is $(0.9486, 0.3162)$.

We have used simple numbers here in order not to distract the reader from the central idea by arithmetical complexity. The Indian astronomers, in the course of determining the coordinates corresponding to each degree on the circle, naturally ran into much heavier arithmetic. They had the skill, the time, and the patience to cope with it. The principles involved are of course exactly the same, whether the numbers are simple or complicated.

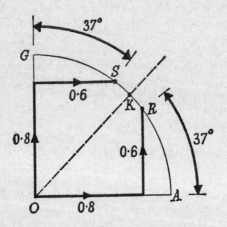

Figure 94

There are certain things that help to reduce the labour. Figure 94 shows a quarter of the unit circle. The dotted line is at 45°. If we imagine the paper folded along this dotted line, the part of the circle GSK will land exactly on the part of the circle ARK. The point R can be reached from O by going 0·8 east and then 0·6 north. The figure shows the corresponding route on the other side of the fold. Starting from O, and going 0·8 north and then

0·6 east we reach the point S, which when folded over would land on R. Suppose somehow we have managed to find out that R is at the position 37°, approximately. The arc GS, which can be folded over to cover the arc AR, must also correspond to an angle of 37°. Since G is at position 90°, S must fall short of 90° by 37°, and so is at 90° − 37° = 53°. Now in going from O to any point, it does not matter whether we go first north and then east, or first east and then north. The amounts will be the same either way. So S is the point (0·6, 0·8). What we have reached here is this; knowing that 37° gives the point (0·8, 0·6) we do not have to make any fresh calculations to find what point is given by 90° − 37°. We simply reverse the order of the two numbers and get (0·6, 0·8) as the coordinates of S at 53°. This halves the work we have to do.

We found earlier that 60° gave the point ($\frac{1}{2}$, $\frac{1}{2}\sqrt{3}$) or (0·5, 0·8660). As 30° = 90° − 60°, it follows that 30° gives (0·8660, 0·5).

We can now see the strategy followed by the Indian astronomers. We know the coordinates of the points at 30° and 0°, so we can find the coordinates at the angle halfway between these, 15°. We found the point at 45° earlier, so we now have the coordinates for 0°, 15°, 30°, and 45°. From these results we can find the coordinates for angles midway between those listed, namely for $7\frac{1}{2}$°, $22\frac{1}{2}$°, and $37\frac{1}{2}$°. This gives us a table in which the difference between each entry and the next is $7\frac{1}{2}$°. Once more calculating the coordinates for angles halfway between those of points already known, we can reduce the intervals between entries to $3\frac{3}{4}$°. The values for points in between can then be estimated with a fair degree of accuracy. I have only discussed angles between 0° and 45°, because those between 45° and 90° can be found by the procedure mentioned earlier, subtracting the angle from 90° and reversing the order of the coordinates.

References to this method seem to be contained in the writings of Aryabhata (born in A.D. 476), who published his mathematics in verse form to aid memorizing.

As has been said, this work has already been done. It is not described here as something you might need to do at some future date. Yet it seemed worth while to report on this classical problem as an example of the power of algebra when applied to an essentially geometrical problem. The algebra has been reasonably elementary. The arithmetic can become tiresome, since all the numbers lie between 0 and 1 and require decimals like 0·8660 or 0·9486 to represent them approximately. Arithmetic is not an insuperable obstacle. In modern times it can be done by electronic computers, in ancient times it could be delegated to an army of assistants. There still remains the problem of telling the computer or the assistants what calculations are to be made. And most people, if confronted with this classical problem and asked to indicate a strategy by which it could be attacked, would have considerable difficulty in finding a way into it. It therefore seemed worth while to show how, at an early stage of mathematical development, it was possible to cope with this problem. The early Indian astronomers did not have the aids we have today – decimal notation, algebraic symbolism, the idea of coordinates. The account above tries to show how these later developments have made the problem, which originally required the utmost ingenuity for its solution, almost a routine matter.

THE DOUBLING PROBLEM

We have just considered the problem of halving an angle. An interesting point arises in the opposite problem. Suppose we know that the point on the unit circle at angle t^o has coordinates (a, b); what will be the coordinates at double that angle, $2t^o$? Figure 95 shows a particular case of this problem. T is the point

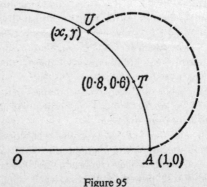

Figure 95

(0·8, 0·6). What is the point U that occurs at double the angle? The broken circle in the figure indicates how we would construct U geometrically; we would put the point of our compasses on T, take TA as radius, and swing round to meet the unit circle at U. The point U thus appears as the intersection of two circles, and this gives us enough information to fix its two unknown coordinates x and y.

Since UT = TA, we have $UT^2 = TA^2$. Here, as usual, we are squaring so as to make use of the distance formula, which now gives us $(x - 0·8)^2 + (y - 0·6)^2 = (0·8 - 1)^2 + (0·6)^2$. This leads to $x^2 + y^2 - 1·6x - 1·2y + 1 = 0·4$. As U is on the unit circle, we can replace $x^2 + y^2$ by 1, which leads us to the equation $1·6x + 1·2y = 1·6$. Dividing by 0·4, we find $4x + 3y = 4$, the equation of a line.

Suggested investigation. Draw the figure on graph paper, taking 5 inches as unit length, so that the point (0·8, 0·6) can be plotted easily. Plot the line $4x + 3y = 4$ just found. What do you notice about it?

* * *

We know that U satisfies $4x + 3y = 4$ and also that $x^2 + y^2 = 1$, since U is on the unit circle. The natural thing to do now would be to solve the first equation either for x or y and substitute in the second equation. As the second equation, $x^2 + y^2 = 1$, contains squares we would expect this to lead to a quadratic, and the solution to require the use of square root tables. However, it turns out to be simpler than this. The reason is as follows. We are seeking the coordinates of U, a point where the broken circle meets the unit circle. However, these circles meet in two points, U and A, and the algebraic equations are not able to make any distinction between these two points. Accordingly we are bound to arrive at two solutions, one corresponding to U, the other to A. If we substitute for y, and obtain a quadratic involving x, one solution of this quadratic will equal the x coordinate of U, the other will be 1, since this is the x coordinate of A. Similarly, if we decided to substitute for x and obtain a quadratic for y, one solution of this would pertain to U, the other solution would be 0, since this is the y coordinate of A.

In the first of these cases, $x = 1$ would be a solution, and so, by the Remainder Theorem, $x - 1$ would be a factor. We could divide by this factor and obtain a linear equation for x. In the second case, $y = 0$ would be a solution, so there would be a factor $y - 0$, that is to say, simply y, and we would divide by y. As it is easier to divide by y than by $x - 1$, it would be better to take the second option than the first. Thus, before we start to make any calculations, simply by thinking about the geometry of the figure, we can decide which would be the better tactic to follow.

The algebra we did earlier showed that the intersections of the two circles lay on the *straight line* $4x + 3y = 4$. This line was the subject of the investigation suggested above. We can now see what this line must be. For the two intersections of the circle are the points U and A. Accordingly both these points must lie on the line $4x + 3y = 4$. That line, then, can be no other than the line UA. In fact it is easy to verify by calculating that A, the point $(1, 0)$, does indeed lie on this line.

In accordance with the tactic decided on earlier, we solve this equation for x. This gives us $x = 1 - \frac{3}{4}y$. On substituting this value in $x^2 + y^2 = 1$, we get $(1 - \frac{3}{4}y)^2 + y^2 = 1$. As we write this equation, we remember our earlier reasoning and think, 'This ought to be satisfied by $y = 0$' (it is) 'and so have the factor y appearing' (as it will). The equation in fact leads to $25y^2 - 24y = 0$, which means either $y = 0$ (corresponding to A) or $y = 24/25 = 0.96$. As $x = 1 - \frac{3}{4}y$, it follows that $x = 0.28$. So U is $(0.28, 0.96)$.

Exercise

It is perfectly possible to carry this work through in the general case, with T the point (a, b). U is then found to be $(1 - 2b^2, 2ab)$. It is instructive to verify this result. Attention must be paid to one fact; as (a, b) is on the unit circle, we must remember that $a^2 + b^2 = 1$. If this is not used at a suitable stage in the argument, an answer will be found more complicated than that given.

SOME ABBREVIATIONS

In most of this chapter we have been concerned with a single problem; to find the coordinates of a point on the unit circle at a position specified by an angle. Both in this work, and in applications of it, we again and again use sentences of the form '*a* is the *x* coordinate of the point at angle t°', '*b* is the *y* coordinate of the point at angle t°'. It is not surprising that it has been found convenient to introduce abbreviations for these frequently used sentences.

The phrase 'the *x* coordinate of the point at angle ...' is abbreviated to the three-letter word 'cos'. The phrase 'the *y* coordinate of the point at angle ...' is abbreviated to the three-letter word 'sin'.

Thus the two statements in the first paragraph of this section can be put in the very compact form, $a = \cos t°$, $b = \sin t°$.

A little practice is needed if we are to become familiar with these abbreviations. Most of the statements given in this chapter provide material for such practice. Thus early on, when we were discussing the regular hexagon, we met a point A with coordinates (1, 0). At what angle does A occur? At 0°. Accordingly $1 = \cos 0°$ and $0 = \sin 0°$. Neither equation here says anything profound. These equations simply tell us the *x* and *y* coordinates of the point at angle 0°. In the same way, the point at the angle 90°, which was shown in Figure 92 as G, is (0, 1). So $\cos 90° = 0$, $\sin 90° = 1$ is the abbreviated way of saying this. At 180° we have the point D of Figure 89 on page 238, with coordinates (−1, 0). So $\cos 180° = −1$ and $\sin 180° = 0$. At 270° we find the point (0, −1), which has not yet featured in any of our diagrams. Accordingly $\cos 270° = 0$, $\sin 270° = −1$.

In this chapter, we have determined the coordinates of various points on the unit circle. All the information so obtained could be expressed concisely by means of *cos* and *sin*.

Exercises

1. In Figure 90, B is at 60°, C at 120°, E at 240°, and F at 300°. The coordinates of B, C, E, F may be read from that figure, or found in the accompanying text. Write this information in the form $\cos 60° = \ldots$, $\sin 60° = \ldots$, $\cos 120° = \ldots$, and so on.

2. We can reach F by going 300° from A, via B, C, D, and E. We can reach F more directly by going 60° in the reverse direction. This is expressed by saying that the point at angle −60° is F. Thus the coordinates of F tell us the values of $\cos(−60°)$ and $\sin(−60°)$.

(i) Write down the values of cos 60° and cos (−60°). What do you notice about them?

(ii) Write down the values of sin 60° and sin (−60°). What do you notice about these?

3. What are cos 45° and sin 45°? (See page 243).

4. The point L in Figure 92 has coordinates $(-1/\sqrt{2}, -1/\sqrt{2})$. State this in terms of cos and sin.

TRIGONOMETRY

Our abbreviations *cos* and *sin* are shortened forms of the words 'cosine' and 'sine'. In fact 'sin 90°', if read aloud, is usually spoken as 'sine 90°'. Otherwise it sounds like a reference to wickedness. These terms belong, of course to trigonometry.

People tend to think of algebra, geometry, and trigonometry as separate subjects, to be kept in watertight compartments. For many years secondary schools in U.S.A. worked by a curious system in which pupils learnt algebra for one whole year, then nothing but geometry for a whole year, and then (if they survived this) trigonometry. How fantastic this procedure was can perhaps be seen from this present chapter, which is entitled 'Algebra as a Key to Geometry', and has in fact been devoted to the question of how, in early times, trigonometrical tables were made. The fact that this could be done without even mentioning the word trigonometry was intended to emphasize the moral that the main thing needed for the mastery of trigonometry is a thorough understanding of algebra.

In this chapter cosine and sine have been defined as the co-ordinates of a point on the unit circle. Anyone familiar with the older books on trigonometry will know that this is not how sine and cosine used to be defined. The older procedure went rather like this. First, a right-angled triangle would be drawn, as in Figure 96. The sides would have lengths a, b, c and the angle indicated would have size, say, t degrees. Then sin $t°$ would be defined as the ratio b/c and cos $t°$ as a/c.

This procedure, as compared with the one used in this chapter, has two disadvantages. First of all, a ratio is a fairly complicated idea. You have to think of the number b and the number c and the process of dividing b by c before you know what sin $t°$ is.

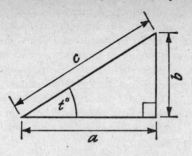

Figure 96

Thus you have to hold three ideas steadily in mind at one and the same time. On the other hand, if you think of sin $t°$ simply as the y coordinate of a particular point on the unit circle, you only have to think of one idea at a time. You find the point in question first. Then you look at the diagram and see at what height (or depth) it occurs. You read off the number, and that gives you sin $t°$ directly. Now the difficulty of learning a subject depends very much on how many ideas have to be grasped simultaneously. There are many more people who can hold one idea in mind than there are who can cope with three ideas at once. This means that more people can learn trigonometry by the coordinate approach than by the ratio approach.

A second difficulty occurs as soon as we come to consider angles which do not lie between 0° and 90°. It is impossible to draw Figure 96 so as to make the angle marked $t°$ of size 120°. Accordingly the teacher has the job of persuading the pupils that the definitions of sine and cosine can be extended to angles of any size. This, as a rule, the pupils do not take kindly to. The stronger mathematicians may accept it quite readily, but many of the others, having learnt to associate sine and cosine with the right-angled triangle in Figure 96 will resist using these terms in any situation where that triangle cannot be drawn. Now one might urge that the idea of extending a definition is an important one. Arithmetic begins with the simple counting numbers, 0, 1, 2, 3, . . ., and then goes on to extend the ideas of addition and multiplication to fractions, negative numbers, irrational numbers, and complex numbers. It could be argued that starting with the definition from the right-angled triangle, and then scrapping this definition in favour of a more general one, is instructive and gives pupils yet another experience of extending a definition. We could perhaps meet this by giving a historical note, that sines and cosines used to be introduced by means of the right-angled triangle and that learners then had to go through the difficulties of accepting a wider definition. But it does seem wise to begin with the coordinate definition, which completely bypasses this difficulty.

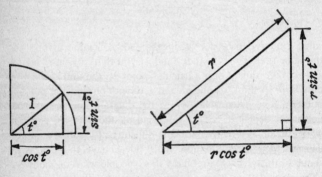

Figure 97

The coordinate definition gives us everything the older definition did. If the angle in question lies between $0°$ and $90°$, the coordinates of the point P on the unit circle are $\cos t°$ and $\sin t°$ as shown at the left of Figure 97. If we enlarge this diagram r times, we obtain the triangle shown at the right. If the sides of this triangle are a, b, c, then $a = r \cos t°$, $b = r \sin t°$ and $c = r$. The ratios b/c and a/c are what they ought to be on the older definition.

THE MALTESE CROSS

Figure 98 resembles a Maltese Cross. Simply by looking at this figure, it is possible to remember a number of facts which learners of trigonometry usually need to know. With the help of

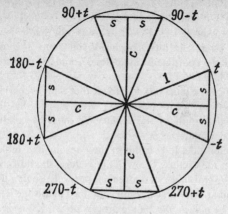

Figure 98

the figure, it requires only a few minutes to become quite familiar with these facts. However, it is by no means unusual to find even university students getting these facts wrong, so it seems worth while to give this diagram and to emphasize the way in which it is used.

The circle in this figure is our usual circle, of radius 1 with centre at the origin. As we go round this circle, starting from 0°, the first point we meet is at angle $t°$ and is marked accordingly. The angles, in degrees, at which the other points occur are also marked. The best thing, of course, is not to look at these printed figures, but to work out for yourself at what angles the other points of the Maltese Cross would occur, so that you can think this diagram out for yourself any time you need it. Finding these angles is quite straightforward.

The point at angle $t°$ has coordinates (cos $t°$, sin $t°$). We can save ourselves writing by shortening these to (c, s). All the corners of the Maltese Cross can be reached from the origin by going a distance c and then a distance s, each in a suitable direction, north, east, south, or west. Accordingly, it is not hard to write down the coordinates of each of these corners. By our definition of cosine and sine, these coordinates tell us the cosine and sine for each of the angles marked. Thus, for example, the point at $(90 - t)°$ has coordinates (s, c). The x coordinate here gives us the cosine, so $\cos (90 - t)° = s$. Similarly, from the y coordinate, $\sin (90 - t)° = c$. This is not really new to us. We had essentially this conclusion on page 247, where we compared the coordinates at angles 37° and 53° (the latter being $90° - 37°$).

Care of course must be taken to see that a minus sign is included where necessary. Thus $(270 + t°)$ gives a point that lies to the east and the south of the origin. We must be careful to record its coordinates as $(s, -c)$.

It does not take long to fill in these results in a table such as appears in the exercise below. This exercise should be done not once but several times, at convenient intervals, so that it becomes quite familiar. Any time a result contained in this table is needed, it should be *thought out* by the process described above. In time, this thinking out will cause the results to be remembered, more or less clearly. But we do not rely on memory; each time we want a result, we make a rapid sketch and think it out again.

The only thing we do have to remember correctly is that the y coordinates gives the *sine* and the x coordinate the *cosine*. If these get mixed up, of course incorrect results will follow.

While it is convenient to work with the short symbols c and s in this exercise, it may on occasion be necessary to write the final result in full. Thus for instance our conclusion four paragraphs ago could be stated as $\cos (90 - t)° = \sin t°$, $\sin (90 - t)° = \cos t°$.

Tables of sines and cosines give values only for angles from 0° to 90°. For angles outside this range, we use the type of consideration we have just been discussing. For instance, if we need $\cos 150°$ we can regard this either as $\cos (180 - 30)°$ or $\cos (90 + 60)°$, whichever we prefer.

Exercise

Complete the table overleaf. Two procedures are possible, and it may be instructive to use both in turn and compare them. One way is to write both coordinates of each point in turn. The other way is to go down the first column, writing in the x coordinates, and then to go down the second column, writing in the y coordinates.

Angle in degrees	Cosine (x coordinate)	Sine (y coordinate)
t	c	s
$90 - t$	s	c
$90 + t$		
$180 - t$		
$180 + t$		
$270 - t$		
$270 + t$	s	$-c$
$-t$		

DESERT ISLAND TRIG. TABLES

Trigonometry has been discussed in various books already published by Penguins, for example in Chapter 13 of *Mathematician's Delight*. Sidelights on trigonometry, some involving calculus and some involving matrices, occur in *Mathematician's Delight*, *Prelude to Mathematics*, and in Chapter 9, 'What is a Rotation?' of *A Path to Modern Mathematics*, Number 4 in the present series. Accordingly, the subject will not be developed further here, except for one final topic, which illustrates an application of the 'trick' given at the beginning of the chapter 'A Method of Discovery'.

The problem is this. You are shipwrecked on a desert island. You have built yourself a boat and have some knowledge of navigation. The only thing lacking is a set of tables, giving sines and cosines. How, using only what you remember from a course in trigonometry taken many years ago, can you improvise these?

Now I presume the navigational instruments you have been able to make for yourself on the desert island are unlikely to have an accuracy as high as one part in a thousand, so tables to three places of decimals should be quite sufficient.

I intend to discuss only how you could make a table giving the sines of angles between 30° and 60°. (Similar principles can be applied to other parts of the tables.) As we have seen earlier in this chapter, the sines of 30°, 45°, and 60° are fairly simple to remember or, if you have forgotten them, to work out. We have in fact $\sin 30° = \frac{1}{2}$, $\sin 45° = 1/\sqrt{2}$, $\sin 60° = \frac{1}{2}\sqrt{3}$. Presumably you do not have tables of square roots with you, but by some standard method, or by trial and error, we suppose you succeed in finding the approximate values $\sqrt{2} = 1\cdot414$, $\sqrt{3} = 1\cdot732$. With the help of these you can put your results in the form $\sin 30° = 0\cdot5$, $\sin 45° = 0\cdot707$, $\sin 60° = 0\cdot866$. Figure 99 shows these three results plotted on graph paper. This figure suggests one way in which we could estimate the sines of angles lying between 30° and 60°. We could make a sweep with a pencil and draw a smooth curve through the three points, and then read off

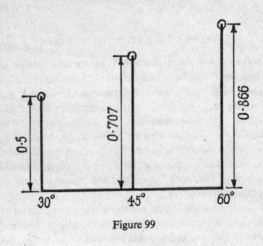

Figure 99

the heights of this curve at intermediate points such as 35°, 40°, and so on. Now of course it is rather difficult to read a graph to three places of decimals, and so we might start thinking whether it was possible to find some numerical procedure that would have the effect of putting a smooth curve through the three points. What kind of curve shall we try to fit to these points? If there were only two points presumably we would join them by a straight line, and fit an equation $y = mx + c$. But our three points do not lie in line. We cannot put a line through them; the graph has to pass through three points, the conditions for this will give three equations, and the line contains only two quantities m and c, not enough (in general) to fit three equations. A circle can be put through three points. We could get a smooth curve through these three points by constructing the circle that passes through them. This however is not very convenient algebraically. Our circle, when found, will have an equation of the form $x^2 + y^2 + 2gx + 2fy + c = 0$. This contains y and y^2, so, if we substituted any particular value of x in which we were interested, we would get a quadratic equation for y, and we would have to solve this in order to find the height of the graph. These difficulties can be avoided if we decide to fit a curve of the form $y = ax^2 + bx + c$ to the three points. This puts three numbers a, b, c at our disposal, which ought to be enough to fit three equations.

There is no suggestion here that the graph of $y = \sin x°$ between $x = 30$ and $x = 60$ is really either a circle or a part of a parabola, $y = ax^2 + bx + c$. The idea is rather that *any* reasonable, smooth curve drawn through the three points will give *more or less* the same values for y. In fact, if you wanted to draw two such curves, say the graph of $y = \sin x°$ itself and the circle through our three points, in order to discuss, say, where one curve was higher than the other and where lower, you would find great difficulty in drawing a plausible diagram. To get any space between the curves and make them appear separate to the eye, you would find yourself bringing in twists and bends which in fact are present in neither curve. From the point of view of exact calculation, there are many quite distinct reasonable, smooth curves through the three points. But for calculations made only to two or three places of decimals, which is what concerns us at present, all these curves are practically indistinguishable. So we choose from them all the one that is most convenient to work with. (I have left vague what I mean by a 'reasonable' curve. Essentially I mean one that is free from bends of the kind you see in a capital S. An S-shaped curve through these points could of course give widely different results.)

The problem we now have, that of fitting an equation $y = ax^2 + bx + c$ to the three points, is the same as the problem at the beginning of the chapter 'A Method of Discovery', to find a quadratic formula to fit a given table. In that chapter we had a trick which gave a simple way of finding the formula, subject only to one condition; the values given must correspond to $x = 0, 1, 2, \ldots$. Now of course the numbers involved here are not 0, 1, and 2, since we have found the sines of 30°, 45°, and 60°.

However, it is quite simple to relate these numbers to 0, 1, and 2, since 30°, 45°, and 60° are spaced evenly apart. Suppose, when we are making our graph, we decide to write 30° next to the origin, 45° just 1 inch to the right, and 60° just 2 inches to the right of the origin. If we now agree that x is to stand, not for the angle, but for the *number of inches at which that angle is marked*, our data now do involve $x = 0, 1, 2$. Once we have found a formula involving this new variable x, there is no difficulty in finding the value of y corresponding to any angle. For instance, since 35° is one-third of the way from 30° to 45°, the angle 35° corresponds to $x = \frac{1}{3}$, and $\frac{1}{3}$ is the number we would substitute in the formula to find our estimate of sin 35°. To estimate sin 31°, we would have to substitute $x = \frac{1}{15}$, and so forth.

If, in Figure 99, you imagine that the distance between each upright line and the next is one inch, then the horizontal distances on that diagram will serve to illustrate the values of x. The values of y are as marked in the figure. (It is convenient to use a larger scale for the y values than for the x values in this diagram.)

The points we have to fit are thus given by the following table in which, as previously, we have also included the rows of differences.

x	0		1		2
y	0·5		0·707		0·866
		0·207		0·159	
			−0·048		

Following the 'conjuror's' procedure, we thus find the formula $y = -0.024x^2 + 0.231x + 0.5$. Care is needed here in finding the coefficient of x. We have to remember that we are subtracting the negative number, -0.024, from 0.207, the first number in the first row of differences, and so obtaining $0.207 + 0.024 = 0.231$. In any work of this kind, it is wise to check the formula obtained, and verify, that for $x = 0, 1, 2$, it does actually give the values of y in the given table.

As explained earlier, to estimate sin 35° we put $x = \frac{1}{3}$ in our formula. This gives, to three places of decimals, 0·574. The printed tables give 0·5736, so our result could not be better. Putting $x = \frac{2}{3}$ gives sin 40° as 0·643, again in complete agreement with the official value, 0·6428. For sin 50° there is an error of

0·001, since $x = 1\frac{1}{3}$ in our formula gives 0·765 as compared with the tables' value 0·7660. For sin 55° we also have an error of 0·001, as $x = 1\frac{2}{3}$ gives 0·818 while the tables have 0·8192. By the standards of a desert island, these results are very satisfactory. Of course, it is not necessary to confine oneself to multiples of 5°. We could equally well find estimates of sin 32°, or sin 37½°, or the sine of any other angle between 30° and 60° that took our fancy.

Now of course our method is most reliable between 30° and 60°, but we might consider how it would work if we went outside that interval. We know that sin 0° = 0 and sin 90° = 1, so we can see the kind of errors that would arise if we pushed the method beyond the interval to which it is best suited. On our graph paper, 0° is marked at $x = -2$. Our formula gives −0·058 instead of 0. For 90°, we put $x = 4$ and obtain 1·040 instead of 1. The errors here, 0·058 and 0·040, are, as expected, much larger than those we had for angles between 30° and 60°.

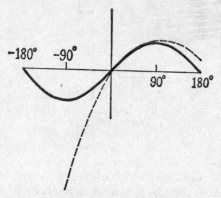

Figure 100

In Figure 100 the continuous curve gives the sine of each angle, the dotted curve gives the estimate found from our quadratic formula. It will be seen that there is an interval in which the two curves cannot be distinguished in a diagram of this size, but that, as we go away from this region, the two curves become wildly different. This is to be expected. We have been using the idea that a small part of one curve is often much like a small part of another curve.* If you are shown an extremely small piece of a circle and an extremely small piece of a straight line, it is difficult to tell which is which. As soon as large pieces are made available, the differences become evident.

The idea that a small piece of a curve may be approximated by a straight line is used in tables. For example, the logarithm of 8·78 is 0·94349, of 8·79 is 0·94399, as given in five-figure tables.

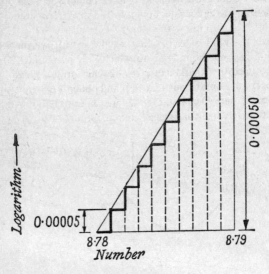

Figure 101

What are we to do if we want the logarithm of 8·783? From 8·78 to 8·79 is a change of only 0·01 which is a fairly small amount. The graph of the logarithm is a well behaved, smooth curve, so we will suppose that a small part of it may be represented, without serious error, by a straight line, as in Figure 101. The difference between 0·94349 and 0·94399 is 0·00050. If we go from 8·78 to 8·781, then to 8·782, and so on, we are going from 8·78 to

*Often, not always. There are extremely wriggly curves, for which no piece, however small, closely approximates a straight line. Such curves, however, rarely occur in elementary work.

8·79 in ten equal steps. With a straight line graph, these ten equal horizontal steps produce ten equal vertical rises, as shown in the figure. As the total rise is 0·00050, the height of each upward step must be 0·00005. In going from 8·78 to 8·783, we are taking three steps to the right; corresponding to this will be three upward steps of 0·00005, making an increase of 0·00015. So we estimate the logarithm of 8·783 as 0·94349 + 0·00015 = 0·94364.

The makers of tables in fact do this arithmetic for us. At the right-hand side of a table of logarithms you will find an arrangement of this kind:

	8	**9**	**1**	**2**	**3**	**4**	**5**	**6**	**7**	**8**	**9**
87 ...	94349	94399	5	10	15	20	25	30	35	40	45

The entries at the right show the heights corresponding to the numbers of steps taken; the decimal points and noughts are omitted. If you consult tables of logarithms, you may not find numbers exactly agreeing with those given here. The numbers have to fit the whole row, and some compromise is necessary.

When computing tables of any kind, we can often reduce the labour by using devices of this kind. A few results can be computed, and then those in between found by means of linear, quadratic, or higher polynomial formulas.

CHAPTER ELEVEN

On Quadratic Equations

In the traditional syllabus, considerable stress was laid on the pupil's ability to solve quadratic equations. In the years 1945–7 I remember discussions at Leicester College of Technology on the place of quadratic equations in life. We were teaching this topic to a great variety of students, some of whom would probably never do more than relatively unskilled work, while others were involved in extremely advanced scientific work. What justification could we offer the students? Who would use it and in what circumstances? Was it, in fact, for some of our students, completely useless, an out-of-date topic that had to be covered only because it still lingered in the prescriptions of some conservative examining body? Such questions are being asked very widely today as the syllabus is subjected to philosophical scrutiny.

The present questioning of the syllabus is to be welcomed. It can easily happen that a mathematical topic is first developed because it is of real and urgent importance in some sphere of advanced work. It gets into the textbooks. The strongest mathematicians, those most likely to become professors and teachers, learn it as children and come to feel that anybody who does not know about it is illiterate. It becomes established as part of the syllabus which everyone has to learn, including those who will never go on to advanced work and find a need to apply it. Centuries can pass without anyone asking, 'When did this get into the syllabus? What purpose did it then serve? Does it still serve that or other purposes today? Who benefits by knowing about it? Should we teach it to all, or some, or none?'

In a matter of this kind there are two extremes to avoid. On the one hand, we do not want to destroy a learner's interest by endless drills on routines that have no apparent purpose. On the other hand, if we are too ruthless in cutting topics out of the syllabus, we run the danger of leaving the learner with such a limited equipment that he cannot read for himself at all, because every book makes references to procedures he has never heard of.

On Quadratic Equations

In examinations candidates are sometimes asked to solve a quadratic equation in which particularly large or awkward numbers occur. This tends to give a somewhat mistaken idea of what we really want to know when we study quadratic equations. For the solution of an isolated equation is done blindly. We do not know in what circumstances the equation arose and this deprives us of one way of telling what would be a reasonable solution.

Someone who is learning algebra in order to read some scientific subject has two objects; one is to recognize certain routine procedures in algebra; the other is to become familiar with certain types of situation.

'Routine' is almost a dirty word. It suggests a human being mindlessly carrying out some task that would be better done by a machine. But the ability to recognize routine situations is very definitely an aspect of intelligence. I have known university students who would attempt to read a mathematics book one line at a time and to consider each step in isolation. They would have done very much better to have stepped back and looked at the problem as whole. Perhaps some scientific question involves five unknowns x, y, z, u, v. Then, as we saw in an earlier chapter, we can as a rule find these five quantities only if we have five pieces of information to give us five equations. We observe that the author of the book has in fact used half a page to assemble the necessary five equations. Next presumably he will strive to eliminate four variables and obtain an equation involving the remaining one. Yes, he has taken the rest of the page to do this. Finally, of course, he uses some procedure appropriate to solving that equation and deducing the values of the four variables that were eliminated. There is an immense economy of effort if a book, or a series of lectures, is considered in this way. The strategy is seen and the details fall into their places. We make no attempt to remember these details. They are routine; if we need to recall or to reproduce the argument at some future time, we will work these details out when we come to them. We will certainly not include them in any summary of the work. But we will be most careful to note any key step where the author has shown ingenuity or done something that would not have occurred to us. These are the points where we might go off the rails; these are the points to isolate, to emphasize, and to remember.

The second aim, becoming familiar with various types of situation, means far more than it seems to. Indeed it embodies the principle that makes the difference between failure and success in the learning or teaching of algebra. When you are familiar with the lay-out of a town, you can move around the town with ease and confidence. To be familiar with a situation in algebra means the ability to think about it without anxiety, to have a feeling for what it is, to be sure when you speak about it that you know what you are talking about. To convey this sense of familiarity is the hardest task in teaching mathematics, for it has an intangible quality. It cannot be written down in the shape of theorems or learned by heart. It is something that grows imperceptibly out of a series of satisfactory experiences.

The making of graphs is one of a number of ways in which the meaning of algebra can be brought out. Every formula, every equation should suggest the thought – how would this appear graphically?

On this approach one would naturally draw a number of graphs. In Chapter 11 of *Vision* we studied a number of graphs given by quadratic formulas, such as $y = x^2$ and $y = x^2 - 2x + 1$, and in the following chapter this was found to aid the teaching of negative numbers. The graph of $y = 10x - x^2$ was found to give an arch, which could in certain circumstances be the path of a ball thrown in the air. The question might arise, 'At what point of the curve is the ball at a height of 16 units?' This asks in effect where the graph meets the line $y = 16$. As Figure 102 shows, there are two such places, one corresponding to $x = 2$, the other to $x = 8$. The ball passes through (2, 16) when it is rising and through (8, 16) when it is falling. There are thus two solutions of the equation $16 = 10x - x^2$. This is something that never happens with a linear equation. An equation such as $2x + 3 = 11$ has the single solution $x = 4$.

It would clearly be unsatisfactory for a student of algebra to meet only linear equations. Even if nothing was said, the experience of repeatedly arriving at a single solution would create an impression in the student's mind that this was the normal situation. Now in fact linear equations are quite exceptional; you are very lucky if an actual problem leads you to one. You are much more likely to land on a quadratic with two solutions, a

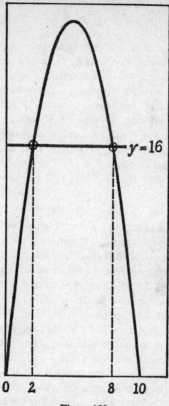

$y = 16$

0 2 8 10

Figure 102

cubic with three solutions, or an equation of still higher degree that may have even more. One would like a learner to be aware of these other possibilities. Now of all these cases the quadratic is the only one that is fairly tractable and lends itself to simple exercises. If an equation contains x^5 or higher powers it is in principle impossible to give a general solution by algebraic means. If the equation contains x^4 or x^3 as the highest power, there is a formula giving the solution, but it is so complicated that it is rarely used and is certainly quite unsuitable for introductory exercises in algebra. Accordingly, if we are to give a learner some taste of what happens when you get away from linear equations, the quadratic seems to be about the only suitable material.

In certain parts of mathematics quadratic equations are particularly likely to occur. For instance, it should be clear from the chapter 'Algebra as a Key to Geometry' that geometrical questions are likely to lead to quadratics. A line is liable to meet a circle in two points; two circles can meet each other in two points. Clearly if we are trying to find such intersections, we should expect that they will be given by quadratic equations, as indeed they are. One can also see this from the fact that the equation of a circle involves $x^2 + y^2$. Consider, for example, the question – where does the line $x + y = 7$ meet the circle $x^2 + y^2 = 25$? If we solve the first equation for y and substitute in the second we get $x^2 + (7 - x)^2 = 25$, a quadratic equation.

The algebraic attack on geometry is important for workers in certain fields. It is probably less important for workers in biology or psychology. Even in those subjects quadratic equations are liable to turn up as incidents in longer calculations. At the end of this chapter an example will be given of the use of quadratic equations in the genetics of human blood types.

Our conclusion then seems to be this; we should not weary pupils to death with long sets of exercises on the numerical solution of quadratic equations involving complicated numbers. We should however use simple examples of quadratics to show that an equation may have more than one solution. Pupils should meet quadratics in a variety of applications, and should become familiar with the way quadratics behave and are handled, so that they are not taken aback if a quadratic equation appears in the course of their own reading or investigations. Actually, quadratic equations have appeared briefly and incidentally a number of times already in this book, without it being necessary to mention the routine by which such equations can be solved. This routine however we will now examine.

SOLVING A QUADRATIC EQUATION

A pictorial method for solving quadratic equations has been known since the times of antiquity. It has certain limitations. In particular, it tends to emphasize one solution of the quadratic and obscure the other. Accordingly, we shall have to consider the method first and later show how to overcome its limitations. Even so, it makes a good starting point, and brings out clearly the main idea and the procedure followed.

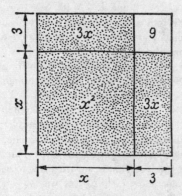

Figure 103

We consider as an example the equation $x^2 + 6x = 40$. Figure 103 shows how we can picture the left-hand side, $x^2 + 6x$. A square of side x will have area x^2. If on each of two sides we put a flap of breadth 3, each flap will have area $3x$. Thus the area of the shaded region in Figure 103 is $x^2 + 6x$. The equation $x^2 + 6x = 40$ tells us that x is to be chosen to make this shaded area equal to 40. Now there is a space in the upper right-hand corner that seems to be calling out to be filled. This unshaded space inside the dotted lines is a square, 3 by 3. Its area is therefore 9. If we combine this square with the shaded region, of area 40, we accordingly obtain a total of area 49. But it is evident from the figure that this combination gives us a square of side $x + 3$. Accordingly $(x + 3)^2 = 49$.

In ancient times, when negative numbers were not considered acceptable, the argument concluded simply that $x + 3$, the side of the square, must be 7, so x must be 4. Today however we have to admit that there are two numbers, 7 and -7, with the square 49.

Accordingly we have two possibilities; either $x + 3$ is 7 or $x + 3$ is -7. The first possibility leads to $x = 4$, the second to $x = -10$. Both numbers do in fact satisfy the equation. If $x = 4$, $x^2 + 6x = 16 + 24 = 40$. If $x = -10$, then $x^2 = 100$ and $6x = -60$ and $100 - 60$ is indeed 40. If we were drawing the graph of $y = x^2 + 6x$ we would get a U-shaped curve that would reach the height 40 at the two places, $x = 4$ and $x = -10$.

Naturally, it is impossible to draw a square of side -7, so it is necessary to remember that there is this second solution which the picture cannot bring out.

The method used here is known as completing the square – evidently a reference to the picture, in which bringing in the piece of area 9 does complete the square of side $x + 3$.

The method of course can be explained purely in terms of algebra without drawing any picture. The idea is to add a number to the left-hand side that will give a perfect square, the square of '$x +$ something'. The only thing is to decide what the something must be. Let us call it a. That is, we are trying to get $(x + a)^2$ on the left-hand side. Now $(x + a)^2$, as we saw on page 245 of *Vision*, is $x^2 + 2ax + a^2$. But adding a number to $x^2 + 6x$ will give $x^2 + 6x + \ldots$. This can only agree with $x^2 + 2ax + a^2$ if $2a = 6$, for otherwise the coefficients of x will be different. Accordingly $a = 3$, and the number added, a^2, must be 9.

To many people, this algebraic argument will not be so convincing and illuminating as the pictorial method. Nevertheless, it is worth while to consider it, as there is still a case to be dealt with, in which drawing a picture is not too straightforward. For suppose we have the equation $x^2 - 6x = 40$ where the left-hand side now involves a subtraction. It is possible to draw a picture by using the device shown on page 248 of *Vision*, but not everyone may find this convenient or helpful. So we come back to the algebraic consideration; what do you have to add to $x^2 - 6x$ to get an exact square? What square, in other words, is of the form $x^2 - 6x + \ldots$? Someone who has digested and remembered the work done on page 244 of *Vision* may produce the answer immediately, by knowing that the only square of this form is $x^2 - 6x + 9$, which is $(x - 3)^2$. Someone not so familiar with the squares may need to argue that $(x - a)^2 = x^2 - 2ax + a^2$; if this is to agree with $x^2 - 6x + \ldots$ we must have $2a = 6$ and again $a = 3$. Alternatively, we could make use of the fact that a is permitted to take negative values; to identify $(x + a)^2$ with $x^2 - 6x + \ldots$, we compare $x^2 + 2ax + a^2$ with $x^2 - 6x + \ldots$. The coefficient of x shows us that we must have $2a = -6$, so $a = -3$. The number we must add is a^2, which is 9. By any of these three methods, we conclude that 9 is the number to add.

The solution would then run as follows. We have to solve $x^2 - 6x = 40$. Add 9 to both sides; this gives $x^2 - 6x + 9 = 49$, which may be written $(x - 3)^2 = 49$. Accordingly we have the two possibilities, $x - 3$ may be 7 or $x - 3$ may be -7. The first gives $x = 10$, the second $x = -4$.

In applications one should not be put off by awkward numbers occurring. Exactly the same method is followed whatever the numbers. If we had to solve $x^2 + 148x = 453$ we could draw a picture on exactly the same lines as Figure 103. We would have to have two flaps whose combined area was $148x$. That is, each flap would have to have area $74x$. Making its width 74 would achieve this. If we had to solve $x^2 + 0 \cdot 36x = 0 \cdot 112$, we would want the two flaps to have a combined area $0 \cdot 36x$. Each flap then should have area $0 \cdot 18x$, and the width $0 \cdot 18$ would ensure this. All kinds of numbers may be encountered; the method remains exactly the same.

In our first example we reached the equation $(x + 3)^2 = 49$ and here of course the numbers had been chosen to make the arithmetic simple; it is very convenient to have 49, since it is exactly the square of 7. Of course in real applications such convenient situations are rare. We must expect to arrive rather at some equation more like $(x + 3)^2 = 5$, from which we conclude that $x + 3$ must be either $\sqrt{5}$ or $-\sqrt{5}$, and we will consult tables if we want a numerical result.

AN APPLICATION TO HUMAN HEREDITY

The reason why equations come into the study of heredity is that there is a difference between the mental and physical characteristics we exhibit to the world and the genes we secretly carry within us. This appears for example when two perfectly healthy parents produce a child with some physical or mental defect, or when two parents of mediocre talents produce a child of outstanding ability.

The equations arise when we try to determine, from what we can observe, the hidden factors at work.

The reason why genes can remain hidden is the following. The body needs to manufacture certain chemicals known as enzymes. Each gene carries instructions for manufacturing a particular enzyme. Normally, each man receives two copies of the recipe, one from his father, one from his mother. It may however happen that one of the parents contributes a gene which instead of containing the recipe is a blank sheet. The man may not show any outward sign of this, since the body can often work quite well with a single copy of the recipe. However, this man carries within himself not the usual complement, a pair of recipes, but one recipe and one blank sheet. When he himself becomes a parent, he will pass on one of these; for each child he begets, it is an even chance whether he hands on the recipe or the blank sheet. We have traced what happens for a man: an exactly similar account holds for a woman.* If it should happen that a man and a woman marry, each having one blank sheet relating to the same enzyme, then there is one chance in four that a child produced by them will receive no recipe at all for the production of that enzyme, but only two blank sheets. This will often be a disadvantage to the child, but in certain circumstances it may be an advantage, just as in snowy regions it is an advantage to an animal *not* to produce pigment.

Usually the proportion of blanks is quite small. For phenylketonuria, a disorder producing mental retardation, about one person in 100 carries a blank. The chance that both partners in a marriage carry a blank is 1 in 100×100, that is, 1 in 10,000. There is then 1 chance in 4 that a child will be unfortunate enough to draw a blank from both parents, so the actual proportion of children born with this disorder is 1 in 40,000. Now the nature of the disorder is understood a treatment by diet is being developed (see C. O. Carter, *Human Heredity*, Penguin Books, page 41).

It is difficult to draw diagrams to illustrate such small fractions. Figure 104 is therefore frankly unrealistic. It shows the effects where 1 gene out of every 4 is blank. The black triangles represent

---

* We are not considering sex-linked genes here.

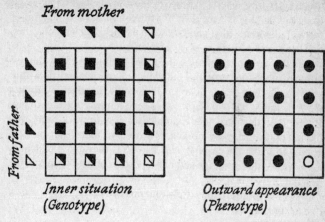

From mother

From father

Inner situation
(*Genotype*)

Outward appearance
(*Phenotype*)

Figure 104

effective recipes, the hollow triangles show blanks. At the conception of the child, it is as if some random process selected a row and a column of the left-hand square. Since only 1 row out of 4 is blank, we have 1 chance in 4 that the father will contribute a blank. Similarly from the columns, there is 1 chance in 4 that the mother will contribute a blank. The square where the row and column cross indicates the genes that the child will carry. A black square indicates that an effective gene has been received from each parent. The squares with half white and half black indicate that the child carries one effective and one blank gene. The white square at the bottom right indicates a child carrying a pair of blank genes. The circles of the right-hand square indicate how the child would appear outwardly. In only 1 of the 16 possible and equally likely cases is the child unable to make the enzyme in question. As a matter of terminology it may be noted that geneticists refer to the outward appearance as the phenotype, while the genes an individual carries determine his genotype.

It will be seen in the left-hand square that there are 9 black squares, 6 partly black, and 1 blank square. This diagram resembles Figure 93 on page 171 of *Vision*, illustrating $(3 + 1)^2$ and leading up to the formulas for $(n + 1)^2$ and $(x + y)^2$. This type of diagram, or the idea it represents, will often be met in genetic literature. More generally, if we had *a* rows of effective genes and *b* rows of blanks, and the same for the columns, we would find a^2 cases of children with a pair of effective genes, $2ab$ with one effective gene, and b^2 with both genes blank.

Usually we cannot observe the genes themselves, but only the outward appearance of the individual. Experimental data thus refer to the right-hand square of Figure 104, in which we do not see 9, 6, and 1, but two classes have been lumped together to give 15 and 1. In this particular example it is easy to work back from the phenotype to the genotype. We observe that 1 out of 16 is unable to make the enzyme, and 1 in 16 is the square of 1 in 4. Thus we conclude that the actual proportion of blank genes in the population is 1 in 4.

However, more complicated situations can occur as is shown by our next example.

HUMAN BLOOD GROUPS

In C. O. Carter's book *Human Heredity*, which was mentioned in our last section, Chapter 4 is devoted to the question of human blood groups. As that chapter will show, this is a question with many facets. We are here concerned with only one of these facets, namely the ABO blood groups, which are a matter of importance in blood transfusion. The data used are based on the book *Outline of Human Genetics* (Heinemann, 1959) by the distinguished geneticist L. S. Penrose. As Figure 105 shows, there are four possibilities for the father's contribution; these are denoted by O, A_1, A_2, and B. The same four possibilities exist for the mother's contribution. If a child receives O from both parents his blood group naturally enough is called O; if A_1 is received from both parents, the child's blood group is called A_1, and so on. But some of the genes seem to be stronger than others. If a child's blood group is A_1, we cannot be sure that both parents gave A_1 genes; the same effect would be produced by A_1 from one parent and O from the other, or by A_1 from one parent and A_2 from the other. This is shown in the diagram by the dotted region which has the shape of a cross. By examining a child's blood we can tell that the child belongs to one of the five compartments that make up this cross, but we cannot tell which one. Similarly there are three black

From mother

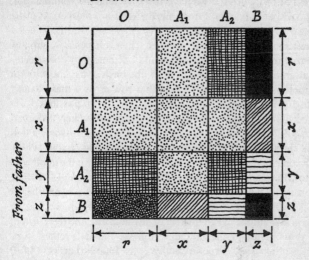

Key to phenotypes

Figure 105

compartments; we cannot distinguish a child who received two genes of type B from a child who received a B and an O. Thus, by examining a large number of children, we can see the total number who belong to the region marked with dots, or with black, or with a check pattern, but we cannot assign them to individual compartments. Given the number of children of each type, the puzzle is to find the proportions of the O, A_1, A_2, and B genes in the population. For the genes do not occur with equal frequency. For instance, O occurs more often than any of the others. Our problem is, to get back from information about the proportions of the various blood groups in the children to a knowledge of the percentages in which the various genes occur.

We will simulate the genetic process in the following way. Suppose that in Figure 105 we have 100 rows and 100 columns. The first r rows are marked O; the next x rows are marked A_1; then come y rows marked A_2, and finally z rows marked B. The columns are divided in the same manner. Then, as earlier, we suppose one row to be selected at random and one column to be selected at random; this is supposed to mark the birth of a child, with the characteristics of the place where that row crosses that column. With 100 rows and 100 columns there are 10,000 possible places inside the big square, and each place is equally likely to occur. Thus, of 10,000 children born, we expect to find as many with A_1 blood group as there are places within the dotted region. Within each compartment of this region, we can write down the number of places. There are x^2 places in the square at the centre of the cross, rx places in the rectangle to the left of that square, and another rx in the rectangle above it; in the rectangle below it we find xy places, and another xy in the rectangle that forms the right arm of the cross. Thus in the dotted region we find altogether $x^2 + 2rx + 2xy$ places. Experimentally it is found that of 10,000 children born, 3,507 belong to this type. Accordingly we can equate the algebraic expression just found to 3,507, and obtain an equation – admittedly, one that does not look as if it is going to be at all helpful! As the square is made up of six regions (corresponding to the six blood types the children may exhibit) we can obtain six equations, one for each region. For example, there are r^2 places in the blank square at the top left corner of the big square, and the number of children observed with blood type corresponding to this square (type O) is 4,356. So we have $r^2 = 4,356$. The black region consists of one square, of side z, and two rectangles each z by r. The number of places in this region is thus $z^2 + 2rz$ which we equate to the number of type B children, namely 828. In this way we obtain the six equations below. These are based on the following figures for the blood types of the children; type O, 4,356; type A_1, 3,507; type A_2, 973; type B, 828, type A_1B, 252; type A_2B, 84.

(i) $r^2 = 4,356$

(ii) $x^2 + 2rx + 2xy = 3,507$

(iii) $y^2 + 2ry = 973$

(iv) $z^2 + 2rz = 828$

(v) $2xz = 252$

(vi) $2yz = 84$

On Quadratic Equations

Now here we have six equations for the four unknowns r, x, y, z so we have a considerable choice as to which equations we use. Equation (i) is very helpful; it immediately gives us $r = 66$. Now we know r and our attention turns to equations (iii) and (iv), since in (iii) the only unknown now is y and in (iv) the only unknown is z. We deal with these by completing the square. We need to have in mind the standard result $(a + b)^2 = a^2 + 2ab + b^2$, in such a way that when we see something resembling $a^2 + 2ab$ we think, 'This wants to have b^2 added to it.' Now we do see such a thing in equation (iii). There we have $y^2 + 2ry$ which wants to have r^2 added to it. From equation (i) we know the numerical value of r^2. Accordingly, adding r^2 to both sides of equation (iii) we find $y^2 + 2ry + r^2 = 973 + r^2 = 973 + 4,356 = 5,329$. So $(y + r)^2 = 5,329$. From a table of squares or square roots we find $y + r = 73$. As $r = 66$, this means $y = 7$. Equation (iv) is handled in exactly the same way. Adding r^2 to both sides leads to $(z + r)^2 = 5,184$, so from tables $z + r = 72$. Since r is 66, z must be 6.

What we have just done can be seen geometrically from Figure 105 by the same idea of 'completing the square'. If the blank square at the top left corner and the three pieces of the region with the check pattern were cut out, they could be fitted together to make a square. We know how many places there are in the checked region and in the blank square; by adding these we can determine how many places there are in the new square just made, blank with a check border. From that we can determine the side of the new square, and then, knowing the blank square has side 66, we can find y, the width of the check border. This is the geometrical counterpart of the algebra we did when we solved equation (iii) for y. Similarly, we can make a square if we border the blank square with the pieces that make up the black region. By applying the same argument we find z in a manner that reflects the algebra used in solving equation (iv).

Now that we know r, y, and z, the simplest way to find x is perhaps to reflect that, since there are 100 rows altogether in Figure 105, x must be the number required to make up that total. As $r = 66$, $y = 7$, and $z = 6$, this means that x must be 21. Alternatively, we could find x by substituting $z = 6$ in equation (v). This would give $12x = 252$ and once again $x = 21$.

We have now used four equations, (i), (iii), (iv), and (v) to find the four unknowns r, x, y, z. We are encouraged in our belief that the underlying theory is sound when we find that the values obtained also satisfy the remaining two equations, (ii) and (vi), since in general it is not possible to find four numbers to satisfy six equations.

This example has been given as an illustration of quadratic equations arising in recent research and exposition. We have not gone into a complication that would certainly arise in experimental work. In actual practice, chance variations would occur, so that in 10,000 cases you would not find the exact numbers predicted by the probabilities. The problem would be to find values of r, x, y, z that fitted the data as well as could be expected. Nor is it likely that all the square roots would come out as neatly as in this example. But neither of these considerations affects the point being made.

CHAPTER TWELVE

Transformations

WHEN we are studying some object, it may help us to observe that it resembles something else which we already know well. The resemblance may be strong or weak. We would agree that the resemblance between a photograph and its enlargement is very strong. The resemblance between a man's outline and his shadow thrown by the setting sun is weaker. Still weaker is the resemblance between the skeleton of a man and a dog. The resemblance is there; man has only the vestiges of a tail and the dog does not feel the need for a big toe, but apart from this, as Sherwood Taylor mentions in his classic, *The World of Science,* the skeleton of a dog corresponds bone for bone to that of a man. The shapes of individual bones differ considerably – those of the head and face, for instance, owing to the different role the brain plays in man and dog, and other bones which are related to upright or horizontal posture.

In this chapter we shall not be concerned with distortions of shape so severe and complicated as those needed to transform a bone of a man into the corresponding bone of a dog. Mathematics has been applied to biological questions of this kind. In D'Arcy Thompson's *On Growth and Form* (Cambridge, 1942) there is a chapter, 'The Comparison of Related Forms' in which simple mathematical procedures are given for changing a drawing of one species of fish into a drawing of another. Here however we shall only be concerned with the simplest operations, such, for example, as that already mentioned, the enlarging of a photograph or a diagram. Our aim is to translate such operations into algebra. Suppose we see two equations; how would we recognize that the graph of one was simply the graph of the other on an enlarged scale? How would we recognize it, if one graph was simply obtained by moving the other a certain distance to the right or left, or a certain distance up or down?

Our problem, in each case, will fall into two parts; first, how do we represent such an operation algebraically, and second, how do we find the effect on a given equation?

The first part, at least, is unexpectedly simple, and can be used to provide interesting exercises for young pupils who know nothing more than how to plot a point on squared paper. For instance, a question could be posed as follows. First plot the three points, A = (2, 3), B = (1, 1), C = (3, 1). Now add 5 to the first coordinate of each point, and plot the resulting points. What do you notice? Figure 106 shows what they would find. When we

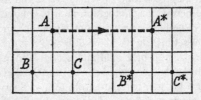

Figure 106

increase the first coordinate in (2, 3) by 5 we get (7, 3). In Figure 106, this point is marked A* to show that it is the point that resulted when we applied our procedure to the point A. We shall often say that the operation 'sends A to A*'. In the same way we find the points B* and C*. It will be seen that the effect of our rule is to shift the triangle ABC by 5 units to the right.

It will be seen that this exercise involves nothing more than applying a simple arithmetical rule, plotting points, and then observing the geometrical effect, which has a certain interest. By working a number of exercises of this kind, one can acquire a feeling for the geometrical meaning of the simpler rules.

Exploratory exercises

What is the effect on the triangle ABC of the following:
1. Increase the second coordinate of each point by 5.
2. Increase both coordinates by 5, so that, for example, (2, 3) is sent to (7, 8).
3. Double both coordinates.
4. Double only the first coordinate.
5. Double only the second coordinate.

6. Subtract the first coordinate from 8, so that, e.g., (2, 3) is sent to (6, 3).
7. Subtract the second coordinate only from 8.
8. Subtract both coordinates from 8, so that, e.g., (2, 3) is sent to (6, 5).

The triangle ABC has a certain symmetry, so that, unless you pay special attention to the letters placed at the corners, you do not readily distinguish between Figure 106, in which the triangle slides to the right, and that for question 6, in which the triangle turning over is involved. This disadvantage can be overcome by considering some shape free from symmetry. The letter F is often used. We can make a capital F by joining the points (1, 1), (1, 2), (1, 3), (2, 2), (2, 3) as shown in Figure 107. In this figure, we also

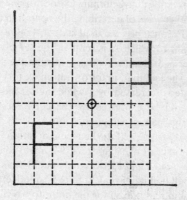

Figure 107

see the new position to which the F goes under the operation specified in question 8 above. The effect of the operation is then clearly seen without the need for distinguishing labels on the five points used. One could describe the effect in various ways. Some would say that the letter F is moved and then rotated through 180°. Others would say that the letter F could be moved from its old to its new position by sticking a pin through the point (4, 4) and then rotating the paper through 180° about this pin.

SPECIFYING OPERATIONS

In the discussion and exercises above, the operations have been described at length in words. The first operation was specified as increasing the first coordinate by 5. This clearly could be put more briefly by using algebra. The first coordinate is x; increasing it by 5 gives $x + 5$. Accordingly we could specify it by saying that (x, y) is to be sent to $(x + 5, y)$. This tells us where each point goes; if we want to know where a particular point is sent, we need only substitute its coordinates for x and y. The operation can be specified completely in terms of symbols as $(x, y) \to (x + 5, y)$, where the arrow is to be read as 'goes to'. (The arrow is used in other branches of mathematics to mean 'tends to' or 'implies'. These other meanings have nothing to do with our present topic.)

There are other ways of describing the operation that are sometimes useful. For instance, we used stars in Figure 106; there A* meant the point to which A is sent. In the same way, we could make an agreement that (x^*, y^*) is to denote the point to which (x, y) is sent. The fact that, in the operation we first discussed, the first coordinate is increased by 5 while the second coordinate is unaltered, would be expressed by the equations $x^* = x + 5$; $y^* = y$, and this pair of equations would specify the operation completely and just as well as the arrow notation.

Sometimes, instead of using stars, we might make use of the distinction between capital letters and small letters. Later in this chapter we shall find it convenient to use capital letters (X, Y) for the original point, and (x, y) for the point to which it goes. With this system, the operation we have been using as an illustration would be specified by the equations $x = X + 5, y = Y$.

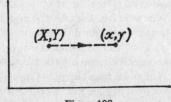

Figure 108

In working with operations, it is very easy to get mixed up between the original point and the point to which it is sent. The wisest thing is to draw a diagram, and to write against the points of the diagram the appropriate symbols. Thus, with the specification last described, the points would be labelled as in Figure 108.

Exercise

Specify the operations in questions 1 to 8 of the previous exercise (on page 280) by means of the system just described, in which the operation sends (X, Y) to (x, y).

TYPES OF TRANSFORMATION

By a transformation we understand any operation that sends each point (X, Y) to a definite point (x, y). Strictly speaking, I suppose we should say, 'a transformation of the plane', for there are other things that could be transformed – for example the earth's surface. If we could arrange for somebody to stand on every place on the globe, and then at a given signal to proceed to another place on the globe, that would (roughly speaking) define a transformation of the earth's surface. Notice that the mathematical meaning is rather different from the everyday meaning. In everyday life we might say a place had been transformed because someone had taken it in hand and made it neat and tidy. This would not be a transformation in the mathematical sense.

It will be evident that a great variety of transformations of the plane exist, if you consider in how many different ways you could give marching orders to the points of the plane. We are considering only a very few and very simple transformations, in fact those of the types already met in the exploratory exercises on page 280.

In the first two questions, each point was displaced the same amount and in the same direction. This type of transformation is technically known as a *translation*. If each point (X, Y) is sent to (x, y) in accordance with the equations $x = X + a$, $y = Y + b$, where a and b are any fixed numbers you like to choose, the effect will be a translation. In question 1, we had $a = 0$, $b = 5$; in question 2, $a = 5$, $b = 5$. It is good to try other values, and see how the translation affects the F on the graph paper, that we had in Figure 107. After trying a variety of values, some positive, some negative, you will come to feel how translations work. How many experiments of this kind you need to make before you feel quite confident and at home with translations will of course depend on your previous mathematical experience.

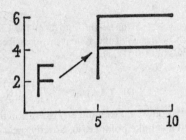

Figure 109

In questions 4, 5, and 6, the effect of the transformation was to change the scale on one or both axes. We shall meet such changes in the chapter on statistics. It is good to observe the effect of a transformation such as (X, Y) to (x, y) with $x = 5X$, $y = 2Y$. This is shown in Figure 109. The new F is twice as tall as the old F, but it may yet appear more squat because the horizontal arms are five times as long as they were. In fact the effect of this transformation is that all vertical lines get doubled in length, while all horizontal lines get stretched to five times their former length. The most general transformation in this family is given by $x = aX$, $y = bY$. It is not necessary for the effect to be that of increasing the length of lines. For instance, with $a = \frac{1}{2}$ and $b = \frac{1}{4}$ we have $x = \frac{1}{2}X$, $y = \frac{1}{4}Y$. This might represent the way in which a very poor piece of material shrunk after being subjected to very severe treatment. The horizontal threads in the weave have shrunk to half their original lengths, the vertical threads to one quarter.

Questions 6, 7, and 8 show that quite simple equations can lead to reflexion or rotation. They simply call attention to this possibility. We shall not follow this theme any further.

TRANSFORMATIONS AND GRAPHS

We have now dealt in some measure with the first part of our undertaking; that is, we have shown how certain algebraic equations are related to certain simple geometrical transformations. We now come to the second part; if a transformation changes one curve into another, how are the equations of the two curves related? By a 'curve' we here mean any graph; it might happen to be a straight line.

We may begin by considering two very simple questions of this type. (i) If the line $y = x$ is drawn and is then moved 5 units upwards, a new line is obtained. What is the equation of this new line? (ii) If the line $y = x$ is moved 5 units to the right, what is the equation of the resulting line?

In a moment both of these questions will be answered, first by a very natural and primitive method, and then by a rather quicker and more sophisticated approach. It is instructive to attack these questions in a common-sense manner before reading on.

*   *   *

The most obvious and natural way to attack such questions is to take a number of points on the line $y = x$, see where they go when the shift occurs, and then try to spot the equation of the line through the new positions. We might take 5 points on the line $y = x$, say $(0, 0)$, $(1, 1)$, $(2, 2)$, $(3, 3)$, $(4, 4)$ and see what happens to them.

For question (i) these points move 5 units upwards. This brings them to the positions $(0, 5)$, $(1, 6)$, $(2, 7)$, $(3, 8)$, $(4, 9)$. We now examine these points to see what equation will fit them. Even without the systematic approach described on pages 308–9 of *Vision* it is easy to see that these points satisfy $y = x + 5$, so that seems to be the equation we are looking for.

For question (ii), shifting the points $(0, 0)$, $(1, 1)$ and so on 5 units to the right gives us the points $(5, 0)$, $(6, 1)$, $(7, 2)$, $(8, 3)$, $(9, 4)$. Here it is evident that the second coordinate, y, is obtained by subtracting 5 from the first coordinate, x, so these points lie on the line $y = x - 5$.

This last answer should be noted, because it is the opposite of the result that a hasty argument suggests. People are liable to reason, 'When we shift 5 units to the right, we add 5 to x. So obviously $y = x$ will change to $y = x + 5$.' This sounds plausible but, as the detailed considerations show, it is the opposite of the truth; the addition in the transformation leads to a subtraction in the equation of the graph.

The method we have just used would be inconvenient if we had to find the effect of a transformation on a circle, such as $x^2 + y^2 = 1$, or a parabola such as $y = x^2$. We might have trouble guessing the equation of the graph through the new positions.

Accordingly we will look at question (ii) by a method that does not involve choosing particular points and guessing an equation. This method is quicker and more efficient, provided you are aware of a possible source of confusion. In our work above, the question

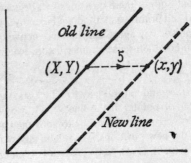

Figure 110

involved the original line $y = x$ and the answer showed that the shifted line had the equation $y = x - 5$. In both equations we had the same symbols, x and y; this makes it very easy to mix up a statement that is true only for points of the original line with one that is true only for points of the shifted line. In order to avoid this confusion, we bring in capital letters X and Y, which will be used in all statements about the old, original line, while x and y will be used for statements about the new, shifted line. Thus the original line can be described as consisting of all points (X, Y) for which $Y = X$. We want to find what equation (x, y) satisfies.

The problem is illustrated in Figure 110. What information do we have about (x, y)? We know that it comes from (X, Y) by a shift of 5 units to the right. What then do we know about (X, Y), the point from which it arises? We know only that (X, Y) is somewhere on the original line. This information is now to be expressed by equations. The first piece of information tells us $x = X + 5$, $y = Y$. The second piece tells us $Y = X$. These three equations embody all the knowledge we have.

Now of course we cannot solve three equations for the four unknowns, x, y, X, Y in such a way as to fix these, nor do we wish to. On geometrical grounds we expect (x, y) to be free to wander up and down the new line; all we hope to find is the equation expressing the fact that (x, y) must stay on that line. This equation should involve x and y only, so our problem is to extract from the three equations an equation containing x and y only, and not mentioning X or Y. This is a problem in elimination, having something in common with the topics considered in the section 'Substituting and Eliminating' (pages 106–110) and in the problems from chemistry. It is by no means hard to carry out. We can solve the first two equations for X and Y; we find $X = x - 5$ and $Y = y$. Substituting in $Y = X$ for X and Y we obtain the equation $y = x - 5$ which our earlier guesswork approach led us to expect.

This method, it will be seen, depends on a somewhat blind manipulation of equations. It is always wise to check the results obtained by it. This may be done by graphing the resulting equation, or also by finding some points that satisfy the original equation, seeing where they go under the transformation, and checking that the shifted points do satisfy the new equation.

Exercises

1. Sketch the line $Y = X + 5$. Find, by the first method, the equation of the line that is produced when this line is shifted 5 units to the right. Also derive this by the second method and check that the results agree.

2. Use both methods to find the effect of shifting the line $Y = X + 1$ through 1 unit to the right.

3. What happens to the line $Y = X$ in the transformation specified by the equations $x = X + 5$, $y = Y + 5$? Is this result reasonable?

4. Find the effect on the line $Y = 2X + 3$ of a shift of 1 unit to the right.

PARABOLAS

In Chapter 11 of *Vision* it was observed that by shifting the parabola $y = x^2$ one could obtain the graphs of $y = x^2 - 2x + 1$ and $y = x^2 - 4x + 4$; indeed many other graphs can be obtained by suitable translations of $y = x^2$. In accordance with the work we have just done, we had better think of the original parabola as $Y = X^2$. Let us consider what happens to this parabola if we apply to it the translation given by $x = X + 2$, $y = Y + 3$; that is, we suppose it to be shifted 2 units to the right and 3 units upward, as illustrated in Figure 111. We will apply the second method, using the first later as a check.

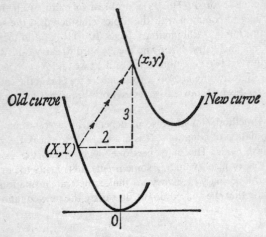

Figure 111

Our three equations have already been written; the equation of the parabola gives us the first, and the two equations for the translation the others. We want to derive an equation in which X and Y do not appear. We can use exactly the same procedure as before; solve for X and Y in the equations of the translation, and then substitute for these in the equation of the old curve. We find $X = x - 2$, $Y = y - 3$ and substituting these in $Y = X^2$ gives

$y - 3 = (x - 2)^2$. When this is multiplied out and simplified it yields $y = x^2 - 4x + 7$ as the equation of the new curve.

We now check this result by considering particular points. The original parabola contained the points $(-1, 1)$, $(0, 0)$, $(1, 1)$. The translation carries these to $(1, 4)$, $(2, 3)$, $(3, 4)$. These do in fact satisfy the equation just found; it is unlikely that we have made a mistake.

It will be seen that the fact that we are considering a curve rather than a straight line does not call for any alteration in our basic method.

Exercises

1. Find the effect on $Y = X^2$ of a shift of 1 unit to the right.
2. Find the effect on $Y = X^2$ of a shift of 2 units to the right.
3. Find the equation of the curve that results when the translation $x = X + 1$, $y = Y + 1$ is applied to the graph $Y = X^2$.
4. Sketch the parabola $y = x^2 - 4x + 5$. Where is its lowest point? What translation would bring $(0, 0)$ to this point? What translation would change the graph $Y = X^2$ into the graph $y = x^2 - 4x + 5$? Your answer to this last question will no doubt be arrived at by examining your sketch. Check its correctness by the algebraic method described above.

THE ELLIPSE

In books on coordinate geometry one can find many theorems about ellipses. It is not obvious to the general reader what relevance, if any, these theorems have to his needs and interests. The ellipse was originally studied by the ancient Greeks as the curve you get if you slice a cone obliquely, and, as the amount of time most of us spend in slicing cones is somewhat limited, one tends to feel that the ellipse is not a fruitful object of study. And indeed it is true that many of the things proved in books about ellipses are of no general interest. And yet the ellipse has a curious history; it plays many more roles in life than its original definition would lead us to expect. One rather specialized application of the ellipse, the fact that the planets move around the sun in ellipses, played a crucial role in the emergence of modern science, since Kepler's discovery provided an excellent foundation for the work of Newton. Kepler had a difficult task since he had to discover the curve in which, say, Mars moves on the basis of observations from the Earth, itself moving in an unknown curve. After great labour he became convinced that one could not fit the observations by supposing the curves to be circles, even though the centres were not at the sun and the paths were described with varying speeds. What then might they be? Presumably some kind of oval, but there are many ovals. It was fortunate for Kepler that the curves were in fact ellipses, something already familiar to him from the work of the Greeks. Otherwise he might have wandered endlessly trying different possibilities.

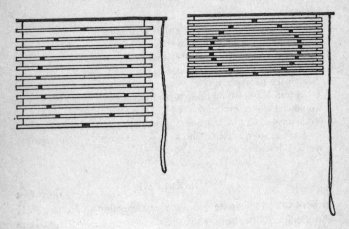

Figure 112

In physics and engineering today there are many studies in which familiarity with certain selected properties of the ellipse is helpful. However our concern at present is not with these, but with statistics, in which a cluster of points having the shape of an ellipse often occurs. Such clusters will be discussed in the next chapter. For this purpose we shall not define an ellipse in terms of sliced cones but rather as a squashed or stretched circle. Imagine a circle drawn on a Venetian blind when the blind is down. Then someone pulls the blind partly up. The result of this is that the various slats come closer together, but there is no change in the horizontal spacing of the points marked (see Figure 112).

Figure 113 shows a way of drawing an ellipse on graph paper. First a circle, centre O, is drawn. Q can be any point on the circle. P is the mid point of MQ. If Q moves round the circle, P will move round an ellipse. This is not a particularly convenient method for actually drawing an ellipse, but it leads very easily to the equation of the ellipse and to some of its properties. Suppose the circle has the equation $X^2 + Y^2 = 1$, and that (x, y) is the point on the ellipse corresponding to (X, Y); that is, (x, y) is the mid point of

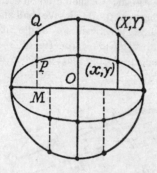

Figure 113

the upright line that ends at (X, Y), as shown in the figure. Here we are dealing with a simple transformation; going from Q to P, or from (X, Y) to (x, y) means that the vertical scale has been halved. We have to write the equations that express this transformation, and then, just as in the other questions we have studied, find what the transformation does to the circle

$$X^2 + Y^2 = 1.$$

* * *

In going from O to (x, y) we go the same distance to the right, but only half as much upwards, as if we were going to (X, Y). This means $x = X$, $y = \frac{1}{2}Y$. These equations specify the transformation.

We next have to combine these equations with the equation $X^2 + Y^2 = 1$ so as to get an equation involving x and y only. It is clear from the transformation equations that $X = x$, $Y = 2y$.

All we have to do is to substitute these in $X^2 + Y^2 = 1$. We obtain $x^2 + 4y^2 = 1$. This is the equation of the ellipse.

In this example we halved the vertical scale. It should be clear that any other fraction could have been used instead of $\frac{1}{2}$ and an ellipse obtained.

A PROPERTY OF THE ELLIPSE

In the chapter on statistics we shall need on occasion to consider changes of scale in the vertical axis. We shall also meet a certain property of the ellipse, which will now be described. We begin by considering a rather obvious property of a circle, and then consider what happens to the diagram when the vertical scale shrinks.

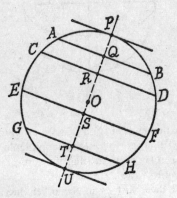

Figure 114

In Figure 114 the dotted line is a diameter of the circle, that is to say, it goes through the centre O. A circle is a very symmetrical figure. If we folded the circle about the dotted line, each point on one side of the circle would land on a point on the other side of the circle; A would land on B, C on D, and so on. As AQ folds over exactly onto QB, Q must be the mid point of AB. Thus the mid point of AB lies on the dotted line; so do the mid points of CD, EF, and GH. Now the lines AB, CD, EF, and GH are all parallel. We could have drawn the figure in a different order, first

the circle, then these parallel lines. Then taking the mid points
Q, R, S, T we would find these lay in line; joining them would
give the dotted line. This then is a theorem about a circle; if you
draw several parallel lines cutting a circle, their mid points lie in
line. You may notice that there are two points, P and U, on the
dotted line that have not been mentioned. These arise from the
parallel lines that just graze the circle. Such a line meets the circle,
as it were, in P and P; the mid point of PP is P, so P must be on
the dotted line, and similarly U must be.

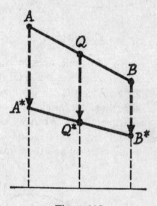

Figure 115

Now what happens to this figure if we shrink the vertical scale?
The circle of course becomes an ellipse. What happens to the
parallel lines? Suppose AB and CD were sent to new positions
and that the two new lines met. We could reverse the process and
come back to our original figure by unshrinking the vertical
scale. The point where the two new lines met would come back
to a point where AB and CD met, and this is impossible since they
do not meet. So the parallel lines must go to parallel lines. It is
also true, as Figure 115 suggests, that mid points are sent to mid-
points. Accordingly, Figure 116 shows the situation after the
vertical shrinkage. Instead of a circle we have an ellipse, but we
still have a series of parallel lines, and their mid points lie on a
dotted line. So the theorem is just as true for an ellipse as for a

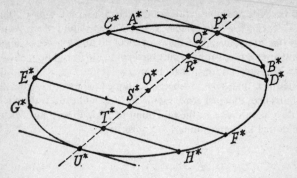

Figure 116

circle; if you draw a series of parallel lines cutting the ellipse, the mid points lie in line. The line goes through O*, the centre of the ellipse, and, just as for a circle, it is called a diameter. In Figure 116, the parallel lines are tilted but the ellipse is in what one tends to regard as its standard position. When we come to use this result in the chapter on statistics, the situation will be as in Figure 117 with the parallel lines vertical and the ellipse tilted. This of

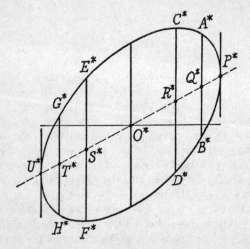

Figure 117

course does not affect the theorem that the mid points will lie in line. This holds for any ellipse in any position and for any set of parallel lines. A geometrical theorem of this kind does not depend on the particular way in which the drawing is placed on the paper. If you care to turn this book in your hands a little, you can bring the ellipse in Figure 117 to its standard position; the parallel lines will then of course appear tilted.

THE EQUATION OF AN ELLIPSE

We shall not have occasion to use it, but in the literature you may occasionally meet the equation of the ellipse with the dimensions *a* and *b* as shown in Figure 118. One way of obtaining this ellipse

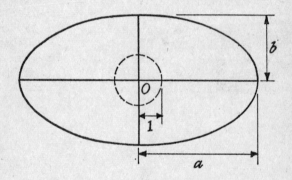

Figure 118

would be to start with the circle centre O and radius 1, and then to enlarge the horizontal scale *a* times and the vertical scale *b* times. This sends (X, Y) to (x, y) where $x = aX$, $y = bY$. As (X, Y) is on the unit circle, we know $X^2 + Y^2 = 1$. By our usual procedure we can eliminate X and Y and obtain an equation involving only x and y. It is customary to leave this equation in the form in which it first arises, and not to get rid of the fractions that appear in it. Of course it would be perfectly possible, and perfectly correct, to multiply it by a quantity that would put it in a form free from fractions, if you wished to do this. However, the form with the fractions left untouched is the one you are most likely to meet in books and papers.

Exercise

By using the three equations just given, find the equation between x and y mentioned above. This is the equation of the ellipse in question.

CHAPTER THIRTEEN

Algebra and Statistics

THIS chapter is most certainly not a quick course on statistics. Indeed any such thing would be most undesirable. In statistics more than in any other branch of mathematics, it is easy to do something that looks entirely reasonable and turns out to be a first-class howler. Another observation on statistics, which is perhaps worth passing on, is that you should not follow the procedure of performing a number of experiments or collecting a certain amount of data and *then* asking a statistician what the results mean. It is quite likely they will mean nothing. The time to consult a competent statistician is before the start of the whole business, so the work is planned to make its results meaningful and the statistical analysis straightforward. This of course does not mean that if, in the course of some work, a striking and entirely unexpected series of coincidences is observed, you should not raise the question whether these are statistically significant. You might miss making a first-rate scientific discovery if you limited yourself by such a severe ban.

The aim of this chapter is simply to illustrate the use of algebra in statistics, and to discuss one or two relevant algebraic results.

Figure 119 is based on data in McNemar's *Psychological Statistics* (Wiley). It shows the results of measuring 192 fathers and sons. In this figure, each dot gives the measurements of a father and his son. For instance, the dot nearest the origin shows a father of height 62 inches with a son of height 64 inches. The highest dot in the diagram shows that in one instance, the father was 71 inches and his son 75 inches.

The question that interests us is – if we know the height of a father, what prediction can we make about the height of his son? How closely are the two measurements connected?

It is clear there is some connexion. A tall father is more likely to have a tall son. In this particular sample, no man 73 inches high has a son less than 70 inches, while no man of height 63 inches has a son as tall as 70 inches. In the diagram the tendency

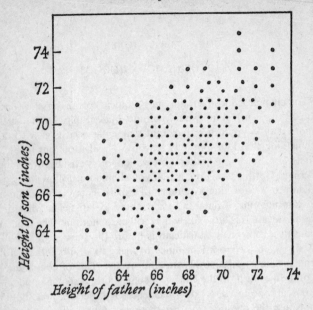

Figure 119

of taller fathers to have taller sons is shown by a tilting of the egg-shaped cluster; the egg points (so far as such an egg can point at all) north-east and south-west. The connexion between the height of a father and a son is limited; given the height of a father, we cannot predict the exact height of his son. For instance, in this sample, fathers with height 68 inches have sons from 65 to 73 inches high.

A much thinner egg is seen in Figure 120, which is based on the following table for the average daily maximum temperature in Aberdeen and Montreal for each month of the year.

Month	Jan.	Feb.	Mar.	Apr.	May	Jun.	Jul.	Aug.	Sep.	Oct.	Nov.	Dec.
Aberdeen	42	40	43	50	55	60	63	63	59	53	47	43
Montreal	26	25	33	42	56	66	73	62	65	54	42	31

Each dot represents a month; its *x* coordinate shows the Aberdeen temperature for that month, its *y* coordinate the Montreal

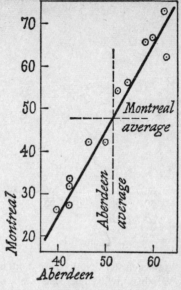

Figure 120

figure. Thus the highest dot in the diagram corresponds to July; its coordinates are (63, 73). The dots cluster rather closely around a straight line. You would have a chance of persuading someone, who did not know how this diagram was arrived at, that the dots represented the results of some laboratory experiment in physics, and that the line drawn in the diagram represented the exact physical law that was being verified. You would not expect to get away with such a story for the diagram of fathers' and sons' heights.

Another thin egg appears in Figure 121, which compares the temperatures in Aberdeen with those in Adelaide. The figures are arrived at in the same way as in the previous example; that is to say, each day the highest temperature is recorded, and then the average is found for each month. This cluster is also long and thin, but there is an important difference. The Aberdeen–Montreal cluster points in a direction between north and north-east; this cluster points more or less south-east.

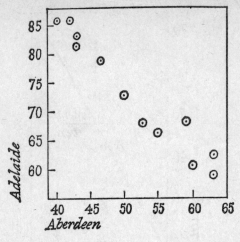

Figure 121

Figure 122 shows a cluster that does not point in any direction at all. The two variables involved here are totally unrelated. They are in fact the temperatures for Aberdeen which we have already used twice, and the initial letters of the months in which those temperatures occur. Thus February, the coldest month, in which the Aberdeen temperature, on the average, reaches only 40° Fahrenheit is shown by the point (40, 6), since F, the initial of February, is the sixth letter of the alphabet. The points, as one would expect, are scattered in a chaotic manner. One would hardly expect to find a connexion between the climate of a month and the position of its name in the dictionary. The letter J, for example, is the initial alike of January, June, and July, one cold month and two of the warmest.

We have now met four diagrams corresponding to four different situations. For the heights of fathers and sons, we have an egg-shaped cluster, indicating a rather loose connexion. For Aberdeen and Montreal, we have a cluster so thin that it is almost a straight line, indicating a tight connexion; if we are told that the temperature in Aberdeen in a certain month is well above the year's average, we can be confident that the same will be true for Montreal. For Aberdeen and Adelaide, we also have a thin

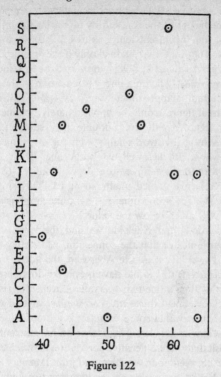

Figure 122

cluster; here again there is a tight connexion, but it is the other way round; the warmer it is in Aberdeen, the cooler it will be in Adelaide. In the diagram for temperatures and the alphabet, we have a random scattering of dots, indicating the absence of any law connecting the two sets of data.

One need not merely speak in terms of fat or thin eggs. It is possible to give a number which measures the fatness or thinness of the cluster, the looseness or tightness of the connexion. This number is known as the *coefficient of correlation* and is usually denoted by the letter *r*. How it is calculated will be described later in this chapter. For the moment we are only concerned with visualizing the kind of situation indicated by various values of *r*. For our first diagram above, $r = 0.56$. Such a value reflects the situation where the height of the father is an important factor in determining the height of the son, but by no means the only one. For Aberdeen and Montreal temperatures, $r = 0.98$, a value very close to 1. The value 1 is obtained only if the dots lie *exactly* on a straight line. A value, like 0.98, lying very close to 1, indicates that the dots very nearly lie on a line. The equation of the line shown in this diagram is approximately $y = 2x - 56$. This means that the Montreal figures could be approximately obtained from the Aberdeen figures by the rule; double, then subtract 56. As the diagram shows, this works fairly well for every month except August, whose dot lies well below the line. Between July and August, Montreal's temperature drops from 73 to 62, while Aberdeen's is maintained unaltered at 63. This single serious departure from an approximate straight line graph is not enough to pull r very much below the value 1.

For Aberdeen and Adelaide we find the value -0.97. The minus sign indicates that the connexion, while close, involves a reversal. The colder it gets in Aberdeen, the hotter in Adelaide. If r had been -1, this would have meant the dots were exactly in line, but this time a line that was falling instead of rising. When r is negative, we speak of *negative correlation*. Anyone who collects evidence to show that more exercise means less illness, or that better teaching means less dependence on punishments, is trying to establish that there is negative correlation between exercise and illness, or between good teaching and punishment.

In our last diagram, comparing temperatures with the alphabetical order of the months, the value of r is -0.04. This value is extremely close to 0, and $r = 0$ indicates, as you might guess, no connexion between the two quantities involved.

The number r is defined in such a way that it can only take values from -1 to $+1$. Figure 123 is meant to give an indication of how the shape of the cluster changes as r goes from -1 to $+1$, in a certain class of situations, of which our first illustration, heights of fathers and sons, was an example.

A common illusion is that a high correlation between two things means that one causes the other. As the weather moves from west to east in this part of the world, it is conceivable that a storm affecting Montreal might later affect Aberdeen. But it is incredible that a warm day in Aberdeen should cause, or be

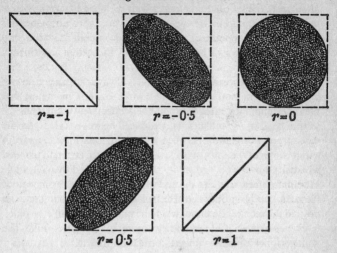

Figure 123

caused by, a cool day in Adelaide. The main reason, I would guess, why there is a strong positive correlation between Aberdeen and Montreal is that both are in the northern hemisphere, and when it is summer in Aberdeen it is summer in Montreal. The strong negative correlation between Aberdeen and Adelaide is because Adelaide is in the southern hemisphere and celebrates Christmas day near the height of its summer.

At first sight, the correlations we found, $+0.98$ and -0.97, seem astonishingly strong. However it must be remembered how the figures were obtained. We are not dealing with a day-to-day comparison. If there is an exceptionally fine day in Aberdeen, the Aberdonians are not justified in saying, 'It must be awful in Adelaide today'. I would imagine much lower correlations would be obtained if figures were compared showing the temperatures at these three places for each day. Aberdeen might have a disappointing day in summer while Montreal had an exceptionally good one. But this would not appear in the tables we have used. In making these tables, the maximum temperature for each day is recorded over a number of years, and the June figure is found by averaging for all the recorded June days. In this way the disappointing days would be merged with the days that were better than expected, and the resulting figure would give no indication whether the disappointing days in Aberdeen coincided with the disappointing days in Montreal or with those that were unexpectedly good.

In considering correlation figures, it is always necessary to know how the information was collected. Mr Lenz, of Christchurch in New Zealand, illustrated this point very effectively in an amusing lecture he gave on statistics. He produced figures which showed a very strong correlation between a boy's success in mathematics and the area of the soles of his feet. The audience began to wonder whether admission to university should be decided by measuring candidates' feet. Right at the end of the lecture he explained how the data had been obtained. He had tested and measured all the boys in the secondary school where he taught. Naturally the older boys had bigger feet and knew more mathematics than the younger ones; that was all the correlation meant.

MEASURING CORRELATION

The number r, the coefficient of correlation, was first considered by Francis Galton, a cousin of Charles Darwin, in work on heredity. Galton was interested in the kind of question suggested by our first illustration: to what extent does the height of a son depend on the height of his father? It is not easy to devise a way of calculating a fraction that will measure this dependence. It may encourage learners who do not see at once why r is defined in the way it is, to know that it took Galton, who was an extremely clever man, fourteen years from the time he first attacked this problem to the time when he found the answer we use today. Galton's first attempt was made in 1875. The final clue to the solution came to Galton in 1889 'during a walk in Naworth Park'. Karl Pearson, at the beginning of Volume 3A of Galton's collected works, comments that when seeking a new idea 'the majority of able men stumble and grope'.

The difficulty in this kind of investigation is that, unlike most experiments in physics, there is a great variation in the observations. A man of height 6 feet may have a son anything from 5 feet 8 inches to 6 feet 1 inch, in the data shown in Figure 119.

A natural way to cope with this difficulty is to ask, 'What is the *average* height of a son, when the father is 6 feet tall?' In Figure 124 we see an oval curve, actually an ellipse, which is a mathematical simplification of the shape of the cluster of dots that represented the fathers' and sons' heights. The small letters, *a*, *b*, *c*, *d*, *e*, *f*, *g* are intended to indicate particular heights that fathers might have. Consider the fathers who have the height *e*.

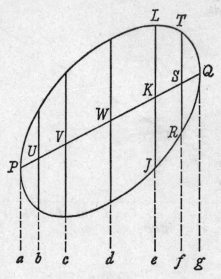

Figure 124

The dots that show the heights of father and son, for fathers in this group, will all lie on the vertical line rising from the letter *e*. (If you have any difficulty with this remark, consider a particular case, say the fathers with height 70 inches. Plot on graph paper the points corresponding to (i) father 70, son 67, (ii) father 70, son 68, and so on until (vi) father 70, son 72. You will notice these all lie in a vertical line. Here we have chosen *e* = 70. We would get similar results with any other number.)

We suppose there are no dots outside the ellipse, so, for fathers with height *e* inches, the lowest dot occurs at J and the highest dot at L. The dot at J corresponds to the shortest son whose father is *e* inches in height, the dot at L to the tallest son with a father of *e* inches height. The average son for such fathers will correspond to the point K midway between J and L. (This argument will be refined somewhat on page 325.)

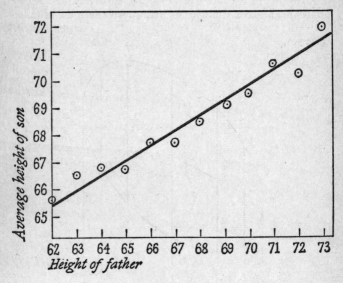

Figure 125

In the same way, we can find the point S which gives the average height for a son whose father is *f* inches in height. We can find other points, such as U, V, W which indicate the average for sons of fathers of heights *b*, *c*, *d* respectively.

You will notice in the figure that the points U, V, W, K, S lie on a straight line. This is in fact a mathematical property of the ellipse, as mentioned in Chapter 12. This line then gives us a graph from which we can read off the average height we expect for a son when we are told the height of the father.

Now of course we get points exactly in line because we have replaced the rather jagged collection of dots in Figure 119 by the smooth ellipse in Figure 124. Figure 125 shows the average height of son for each height of father calculated from the actual data. It will be seen that the points do lie reasonably close to a line. An encouraging feature is that the agreement is best near the middle of the diagram, where the majority of the cases lie. There were twenty-six fathers of height 67 inches, but only five of height 73 inches, and only six of height 63 inches. With only five men to represent all the men of height 73 inches, you are prepared for the possibility that these five cases may not give you an exact indication of the real situation.

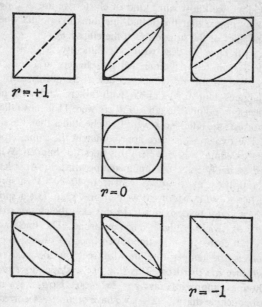

Figure 126

Now the slope of the line in Figure 125 is 0·56 and this is how Galton first arrived at the idea of r; he thought of r as being the slope of the line obtained in this way.

This fits in well with the idea explained earlier, that r measures whether we are dealing with a fat or thin cluster of dots. Figure 126 shows a whole series of clusters, with r going gradually from +1 to −1. We begin with an exact line. Then we have a thin ellipse, which gets fatter until it becomes a circle. Then it gradually gets thinner again, and ends by being once more a straight line. With each ellipse, we see the dotted line that corresponds to the line PUVWKSQ in Figure 124, the line that gave the average height of a son for a given height of a father. You will notice that as we go from left to right, the dotted line swings steadily round. At the beginning, the slope of this line is +1. It has fallen to 0 by the time we reach the circle, and to −1 by the end of the series. Each slope occurs once only; if we know the slope of the dotted line, we know what kind of cluster we are dealing with. You may notice that the dotted line joins the points where the upright sides of the square touch the ellipse.

We still have to deal with one difficulty, a difficulty that bothered Galton for a time. All the clusters in Figure 126 fit neatly inside a square. Now of course there are cases in which this happens. Our first example, with fathers' and sons' heights, was a case in point. The tallest fathers were 11 inches taller than the shortest; the tallest sons were 12 inches taller than the shortest sons. The two spreads were approximately the same. However, there are definitely cases where this does not happen. When we looked at the average maximum temperatures in Aberdeen and Montreal, those at Aberdeen went from 40° to 63°, a spread of 23°, while those in Montreal went from 25° to 73°, a spread of 48° – just about twice as much. In this case the cluster would fit into a box twice as high as it was broad. The line drawn in Figure 120 corresponds to the dotted lines we discussed earlier; it predicts the temperature in Montreal on the basis of the temperature in Aberdeen. This line has a slope close to 2; its equation is approximately $y = 2x - 56$. Now we certainly cannot conclude that $r = 2$. As we have seen, $r = 1$ corresponds to the tightest possible connexion between two sets of readings. Galton found a solution to this difficulty, which may be explained as follows; if the box in which the ellipse sits is not a square, squash it down until it becomes a square. In the Aberdeen-Montreal case, we can achieve this by halving the vertical scale. If we mark the degrees of temperature on the upright axis only half as far apart as those on the horizontal axis, the cluster will then fit nicely into a square, and we can *then* measure the slope of the line and get the value of r.

All our reasoning so far has been pictorial, and reasoning often is pictorial when a man is trying to discover a new idea. For actual computations it would not be at all convenient to estimate from the appearance of a cluster what ellipse it resembled, then to enclose this ellipse in a rectangle and alter one scale so as to turn the rectangle into a square, and finally to draw the dotted line and measure its slope. For such work it is time to turn to coordinate geometry and translate all our pictures into algebra and arithmetic. In order to do this we shall have to make rather more definite and clear some ideas which we have been using rather loosely and vaguely.

MEASURING SPREAD

A little earlier we mentioned that the heights of sons were spread out about as much as the heights of fathers, and that temperatures were spread out in Montreal about twice as much as in Aberdeen. We justified these simply by looking at the tallest and shortest men recorded and at the highest and lowest temperatures. However, this is not a very satisfactory procedure. For instance, there are some very rare cases where a man is fantastically tall or unusually short. It is purely a matter of chance whether such an exceptional man gets included in the sample or not. It might happen that a man seven feet high got included with the sample of fathers chosen. Perhaps this man married a rather short woman, and his son has a height which is not at all unusual. If we simply look at the tallest and shortest heights recorded, we shall be led to conclude that fathers' heights are much more spread out than sons'. On the other hand, if such an exceptional man turned up among the sons, his father being tall but not exceptionally so, we should be led to the reverse conclusion, that sons' heights varied much more than fathers'. Our computations would give widely different results in these two cases. Now it is extremely undesirable that an investigation involving some 200 cases should be at the mercy of an exceptional situation in one of the cases included. Accordingly we look for some sort of averaging procedure, so that the result will depend on all the cases studied, and not merely on the extreme ones, which anyway are the most uncertain and unpredictable.

The Search for Pattern

There are various ways of measuring spread. Suppose for example ten men, A, B, C, D, E, F, G, H, I, and J enter some contest and score respectively 5, 8, 8, 10, 10, 10, 10, 11, 12, and 16. Their scores total 100, so their average score is 10. What interests us at the moment is not their average score, but the way they spread out from the average. Four men, D, E, F, G, conform exactly to the average; their deviation is 0. H is the only man who differs from the average by 1. Three men, B, C, and I, differ from the average by 2. We lump them together; we are not concerned that B and C are below average while I is above average. Finally A deviates from the average by 5 while J differs by 6.

One way of measuring spread is simply to take the average of the deviations listed above. Our list above shows the following:

Deviation	0	1	2	5	6
Number of men	4	1	3	1	1

If you liked you could write the list of deviations out in full; 0, 0, 0, 0, 1, 2, 2, 2, 5, 6. The average deviation is thus 1·8.

The men B, C, and I differ from the average by just about this amount. They could claim that they are not complete conformists, like D, E, F, and G, nor uncomfortably exceptional beings like A and J. They are exhibiting an average degree of nonconformity. The average deviation, 1·8, would give us one way of measuring how much these scores were spread out.

Now there are certain inconveniences in the method just described, particularly when any such situation has to be treated by algebra. This is due to the fact that, in finding the deviation of any individual, we ignore whether he is deviating above or below the average. If a denotes the average and x denotes his score, his deviation is $x - a$ if he is above average and $a - x$ if he is below average, for these are the ways the difference between his score and the average must be written if we are to get a positive number each time. Now it is very inconvenient in an algebraic formula if you have to make distinctions of this kind. One can work with such things; there is a special sign $|x - a|$ which denotes the difference between x and a, but this is not the easiest thing to handle.

Our object is to avoid a formula that will make it appear as if opposite types of nonconformity cancelled out. We do not want J's 6 above average to combine with A's 5 below average by simple addition of $+6$ and -5 to give $+1$. By being extreme in opposite directions A and J are giving variety to society; they are both contributing positively to spread things out. So far as creating diversity is concerned, A and J are cooperating. We have to find some formula that will bring this out, in spite of the opposite signs in $+6$ and -5.

Now in algebra there is a simple operation that always leads to the same sign, namely squaring. Whether a number is positive or negative, its square will be positive. If we add, not $+6$ and -5, but $(+6)^2$ and $(-5)^2$ we get $36 + 25$, and the two parts are pulling in the same direction. Accordingly, we can avoid any possibility of extremes cancelling out if we regard a man who scores x when the average score is a as making a contribution to nonconformity, not of $x - a$, but of $(x - a)^2$.

For our ten contestants the computation would run as follows:

Score, x	5	8	8	10	10	10	10	11	12	16
Average, a	10	10	10	10	10	10	10	10	10	10
$x - a$	-5	-2	-2	0	0	0	0	$+1$	$+2$	$+6$
Contribution, $(x - a)^2$	25	4	4	0	0	0	0	1	4	36

Adding the numbers in the bottom row, we find the total contribution to be 74. The average contribution per man is thus 7·4. Now 7·4 is roughly the square of 2·7, so on this method of accounting, we would have to regard as a man with a typical degree of nonconformity one who scored about 2·7 above or below the average. This number, 2·7 in this example, is referred to as the *standard deviation* and is usually denoted by the letter s or the corresponding Greek letter σ (sigma). This number, σ, is very often used to measure how widely data are spread out. It is particularly convenient for calculations and is the only measure of spread we shall use in the remainder of this chapter. Galton was familiar with it.

In our first example in this chapter, the value of σ for the fathers' heights was 2·49, for the sons' heights 2·33. These numbers are approximately equal. In Figure 119, the x coordinates represented the heights of the fathers, while the y coordinates referred to the sons. For this reason the symbols σ_x and σ_y could be used to indicate the numbers that measure the spread of the fathers' heights and the spread of the sons' heights. We would write $\sigma_x = 2.49$, $\sigma_y = 2.33$. In Figure 120 the x coordinates represent Aberdeen temperatures, the y coordinates those for Montreal. For these data, it can be calculated that $\sigma_x = 8.15$, $\sigma_y = 16.16$. It will be seen that the spread for Montreal is about twice that for Aberdeen, in agreement with the more rough and ready argument we used earlier. Then, you may remember, we saw that, to get the cluster to fit into a square, we had to use a scale on the y-axis about half that used on the x-axis. We could get the same effect by using equal scales but halving the y coordinates. For numerical work this is a more convenient procedure. We halve the values because 8.15, the standard deviation of the x values, is roughly half of 16.16, the standard deviation of the y values. If we wanted to be more precise, we would multiply the y values by 8.15/16.16, which is σ_x/σ_y. This last fraction is one we shall need in the argument later on; in Galton's approach, the value of r is given by the slope of the dotted line only when the cluster has been made to fit into a square, and this is achieved by multiplying all the y values by σ_x/σ_y.

FITTING THE LINE

We now consider the problem of finding the dotted line that runs through the middle of the cluster of dots. This problem is related to a question that occurs frequently in experimental work. We perform some experiments and have reason to believe that these should lead to a straight-line law. In fact, when we plot our points, we find that they do lie more or less in line, but of course experimental errors always occur and we cannot put a straight line exactly through all the points. The question then is – how shall we draw a line that fits these points as well as possible?

Figure 127 shows a way of visualizing this question in terms of a mechanical device – a device that might not be too easy to set up in actual practice. For our line we use a steel rod. Rubber bands pass round this rod, and each band is trying to pull the rod towards one of the plotted points. Things must be arranged in such a way that, if the rod actually passed through one of the points, that point's band would 'feel satisfied' – there would be no ten-

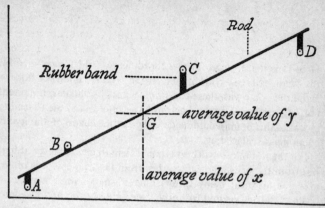

Figure 127

sion in it. But the further away the rod is, the greater the tension in the band must be; in fact the tension must be proportional to the amount by which the rod misses the point. Things must be so arranged that the bands are compelled to remain upright, as shown in the figure. Each band then is doing its best for the point to which it belongs, and under all these conflicting pulls the rod would eventually come to rest in a position which represented a fair compromise.

This mechanical arrangement is equivalent to what is usually called the Method of Least Squares. Here I will not discuss the method but confine myself to reporting the result. To avoid long expressions or troublesome notation, we will keep to the case shown in Figure 127, where there are only four points to be fitted. (The general result can easily be guessed from the result for this particular case.)

The result is in two parts. The first part tells us that the line is bound to pass through a particular point; the second part tells us the slope of the line.

The first part says that when the rod settles down it will pass through a point that we might call the *average point*. The meaning of this can be brought out by a numerical example. Suppose the points A, B, C, D in Figure 127 have the coordinates (2, 1), (6, 5), (18, 13), (30, 15) respectively. We make the following table:

	A	B	C	D	Average
x	2	6	18	30	14
y	1	5	13	15	$8\frac{1}{2}$

Here 14 is the average of the x coordinates of the four points, $8\frac{1}{2}$ the average of the y coordinates. By the 'average point' we understand the point with these two numbers as coordinates, the point $(14, 8\frac{1}{2})$, marked G on the diagram. (The letter G is chosen because in statics this point would be described as the centre of gravity of equal masses placed at A, B, C, D.)

For the rest of our work; it is convenient to measure everything not from the original origin O, but from G. This device is often used in statistics; many results take a simpler form when coordinates are measured from the average point, G, and it is important to note when a formula holds true only if this is done. We will, for the remainder of this chapter, use capital letters for coordinates measured from G. Thus (X_3, Y_3) will denote the coordinates of C measured from G, namely $(4, 4\frac{1}{2})$. Similarly (X_2, Y_2) will denote the coordinates of B measured from G, namely $(-8, -3\frac{1}{2})$. It is essential to preserve these minus signs. The following table gives the (X, Y) coordinates.

	A	B	C	D
X	-12	-8	4	16
Y	$-7\frac{1}{2}$	$-3\frac{1}{2}$	$4\frac{1}{2}$	$6\frac{1}{2}$

Calculating these is quite easy; we simply subtract 14 from the x values in our first table and subtract $8\frac{1}{2}$ from the y values.

We need one more table containing three rows, and then our calculation is nearly complete. In this last table we show for each point the value of X^2, Y^2, and XY, and we find the averages of these:

	A	B	C	D	Average
X^2	144	64	16	256	120
Y^2	$56\frac{1}{4}$	$12\frac{1}{4}$	$20\frac{1}{4}$	$42\frac{1}{4}$	$32\frac{3}{4}$
XY	90	28	18	104	60

Here, under A, we find first 144 because that is the square of -12, then $56\frac{1}{4}$ because that is the square of $-7\frac{1}{2}$, and finally 90, because that is what you get when you multiply -12 by $-7\frac{1}{2}$. Similar remarks apply to the entries under the other letters.

To find the slope of the line in Figure 127 we need only the top and bottom lines of this last table. The slope is found by dividing the average value of XY, namely 60, by the average value of X^2, namely 120. The slope is thus $\frac{1}{2}$.

Galton, you may remember, identified the correlation coefficient, r, with the slope of the line. However we must not jump to the conclusion that r is $\frac{1}{2}$ for the little cluster of points A, B, C, D. Indeed this conclusion looks wrong. In statistics we would hardly draw conclusions from so few points as four; the number was kept so low only in order to avoid long arithmetical calculations. But even admitting this, the cluster A, B, C, D looks far more like the Aberdeen–Montreal diagram, where r was nearly 1, than the well-rounded egg for fathers' and sons' heights, where r was roughly $\frac{1}{2}$. And in fact there is one consideration still to be taken into account. Galton's identification of the slope with r works only when the scales have been adjusted to make the cluster fit into a square, and here again the look of the figure helps us. Figure 127 is certainly nowhere near as tall as it is broad.

At the end of the last section we saw that, to obtain a square, it was necessary to multiply the vertical measurements by σ_x/σ_y. What are σ_x and σ_y in our example? Earlier we agreed to measure nonconformity by the square of the distance from the average. The number σ_x was defined as the distance an individual would have to be from the average for his nonconformity to represent fairly that of the group. Now his nonconformity would be measured by the square of his distance, that is σ_x^2. To be representative, this would have to equal the average of X^2, since the numbers X represent differences from the average. Accordingly we have the equation

$$\sigma_x^2 = \text{the average of } X^2$$

In our example, the average of X^2 is 120, so a table of square roots tells us that σ_x must be 10·95. Similarly we find that σ_y^2 must equal the average of Y^2, namely $32\frac{3}{4}$, so σ_y must be 5·72. (We now see why it was necessary to include the middle row of our last table.)

The appearance of the Figure 127 confirms these results. Comparing 5·72 and 10·95, we see that σ_y is a little more than half of

σ_x, and this looks right. The vertical spread of the points in Figure 127 does appear to be about half the horizontal spread.

Accordingly the vertical scale of that figure needs to be enlarged in the ratio $10.95/5.72$, which is 1.91. Fortunately we do not have to go back and multiply each individual y coordinate by 1.91, for we are only interested in the slope of the line, and such a change of scale will multiply its slope by 1.91. Accordingly, after the change of scale, the slope of the line will be not $\frac{1}{2}$ but $\frac{1}{2} \times 1.91$, or approximately 0.95. This is the value of r and now we have reached the sort of value we should expect for such a thin cluster of points.

The routine just described may strike you as complicated. You will find if you work a few exercises on it that it soon becomes familiar. In the exercises below, the numbers have been 'cooked' so that the arithmetic is not too painful. To make sure that you can think about what you are doing, and do not get exhausted by the mere arithmetic, certain aids should always be used – a table of squares and square roots, a slide rule for the divisions, and some kind of adding device. Not everyone can get hold of a desk calculating-machine; there are inexpensive little flat adding devices which are quite sufficient for this type of work.

The exercises contain instructions which will guide you through the routine. After the exercises, some points involving algebra will be discussed.

Exercises

1. (i) Plot the following four points on graph paper; $(-19, -15)$, $(-9, -10)$, $(11, 6)$, $(21, 27)$. Do they form a fat or thin cluster? Does the correlation appear to be positive, zero, or negative? Roughly what value of r would you expect by looking at the cluster?

(ii) What are the coordinates of the average point, G?

(iii) Make the table showing the coordinates (X, Y) when we measure from G. Check from the graph paper that these have been correctly found.

(iv) What is the total sum, $X_1 + X_2 + X_3 + X_4$, of the X coordinates (measured from G)?

(v) What is the total sum, $Y_1 + Y_2 + Y_3 + Y_4$, of the Y coordinates?

(vi) Tabulate the values of X^2, Y^2, XY. What are the averages for these?

(vii) The slope of the required line equals the average of XY divided by the average of X^2. What is this slope? Draw the line through G having this slope, and check that its position is a reasonable one.

(viii) What are the lengths of the four upright 'rubber bands' joining the given points to this line? Record these with a plus sign if the point is above the line, a minus sign if the point is below the line. What is the sum of the four numbers so obtained?

(ix) What are σ_x and σ_y?

(x) To slide rule accuracy, what is σ_x/σ_y?

(xi) In order to find r, we have to work out what the slope of the line, found in part (vii), becomes when the vertical scale is multiplied by σ_x/σ_y, found in part (x). What is r? Compare your answer with the prediction you made in part (i).

Note. Questions (iv) (v), and (viii) are not essential for the calculation of r. However, they bring out facts of which it is good to be aware, and also serve to check the accuracy of the computation.

2. Answer the same sequence of questions for the points (0, 1), (4, 0), (16, 11), (20, 8).

3. The same questions for (0, 5), (2, −3), (6, 1), (8, 13).

4. The same for (0, 11), (4, 2), (8, 0), (12, 11).

SIMPLIFYING CALCULATION

In the above exercises, the figures were carefully chosen to make the average values of x and y whole numbers. In actual practice of course things are quite different, and awkward numbers are liable to arise. In a high-powered establishment this does not matter. The data are simply fed into a computer, and no one minds how many places of decimals there are. An individual working on his own can buy computer time at a very reasonable rate. Still presumably for a few years yet there will be people working out correlations without the help of an electronic computer, so it seems worth while to mention certain devices which lighten the labour of calculation. Another reason for mentioning these is that books on psychology, ecology, and other applications of statistics, often use these devices in their illustrative examples; to follow the exposition, one needs to understand what is being done.

The temperatures for Montreal, given in an earlier example, show how complexity enters the arithmetic. All these temperatures were given by whole numbers but their average was 47·92. In the routine used in the exercises, the first step is to replace the original numbers x, y by X, Y. which are distances from the average. Thus the January figure of 42 leads to an entry in the X, Y table of -5.92, since 42 is 5·92 below the average value 47·92. The July figure of 73 leads to an entry of 25·08. If you reflect that numbers of this type are liable to occur in both rows of the X, Y table, you can see that the business of finding the average values of X^2, Y^2, and XY may be formidable for someone with limited mechanical aids.

There are two algebraic results that enable us to cut down the labour very considerably. Both depend on the facts suggested by questions (iv) and (v) in the exercises above, that you get 0 if you add together all the numbers in the X row, and similarly for those in the Y row.

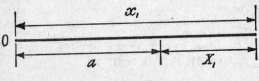

Figure 128

It may be as well to begin by showing that this always happens and that it is not merely something we have guessed. Consider first the simplest imaginable case when we have just two numbers x_1 and x_2. To find their average, a, we add them and divide by 2. Thus $a = \frac{1}{2}(x_1 + x_2)$. It is convenient to multiply by 2 so we have $2a = x_1 + x_2$. Now X_1 is the number you get if you measure from a to x_1. This means, as Figure 128 illustrates, that $x_1 = a + X_1$. (If it should happen that x_1 lies to the left of a, then X_1 will be negative; the same equation still holds good. You may check this from the figures in any of the examples or exercises in this chapter). In the same way, $x_2 = a + X_2$. If we now substitute in the equation found earlier, $2a = x_1 + x_2$, we get $2a = 2a + X_1 + X_2$. Subtracting $2a$ from each side, we get $0 = X_1 + X_2$.

This proves what we want for the case of just two entries. Only very minor amendments are needed to adapt this proof to the case of three or four or five or any other number of entries. For instance, when there are three entries, the average naturally is one-third rather than one-half of the total. It is instructive to write the proof out for the case of three entries, then perhaps for four, and so on until it becomes clear that essentially the same proof will work for any number of entries.

The result, that the sum of the X is always 0, may appear in a textbook in the form $\Sigma X = 0$, and the proof may well be in a very condensed form involving the summation symbol Σ. To many people, expressions involving this symbol do not readily convey their meaning; the way to cope with such expressions is to do what we have done above – go back to particular simple cases and write the thing out in full until you become convinced of the general result.

We need not prove separately that the total of the Y values is also 0. What we have just proved applies to the averaging of any collection of numbers. It does not matter whether they are in a row labelled X or a row labelled Y.

We now come to the question of finding the average values of X^2, Y^2, and XY, the numbers used in finding the slope of the line and the coefficient of correlation r. As we have seen, the numbers X, Y may be arithmetically more complicated than the original numbers x, y. So the question naturally arises; can we make calculations based on the original numbers x, y and then, at the end, make some correction to allow for the fact that these numbers are not measured from the average point (a, b)? In other words, how are the average values of x^2, y^2, and xy related to the average values of X^2, Y^2, and XY?

Most learners of algebra would, I believe, feel rather disheartened if they were asked to attack this question unaided. Yet its solution does not call for anything particularly profound. Let us consider what the difficulties are. The first example of this chapter involved 192 pieces of information. Now naturally, it is awkward to make algebraic calculations involving 192 different symbols. We have already seen how to meet this difficulty – to work first with two symbols, then with three and so on, and to observe that the essential idea of the work does not depend at all on how many quantities are involved. When we work with only two or three quantities, naturally our expressions become much shorter and easier to handle. Most of what we have to do involves only the early routines of algebra – multiplying out a product, squaring a sum, collecting together terms that contain the same symbol. There is perhaps one stage where a certain amount of judgement is required as to how the terms should be collected together. If the collection is done in a certain way, we are able to make use of our earlier observation that the sum of the X values is 0. Someone who is beginning algebra might find it instructive, and helpful to confidence, to attack this problem in the case where only two points are involved, (x_1, y_1) and (x_2, y_2). These should be expressed in terms of the average values, a, b and the measurements X, Y from the average point (a, b). How, in terms of these quantities, do the average values of x^2, y^2, and xy appear?

*   *   *

In this case, the average value of x^2 is $\frac{1}{2}(x_1{}^2 + x_2{}^2)$. As we saw earlier $x_1 = a + X_1$ and $x_2 = a + X_2$. We need to square these and then substitute the results. This gives us, for the average of x^2, the moderately long expression $\frac{1}{2}(a^2 + 2aX_1 + X_1{}^2 + a^2 + 2aX_2 + X_2{}^2)$. The only question now is how best to group the terms inside the bracket. Things work out best if we arrange the terms according to the power of a in each. Inside the bracket we see $2a^2$; no difficulty here. The terms completely free of a are $X_1{}^2 + X_2{}^2$; there is no possible way of simplifying these; we can only record them as they stand. All that remains is now $2aX_1 + 2aX_2$ which may be written $2a(X_1 + X_2)$. But we noticed that the total of the X values is always 0. So $X_1 + X_2$ is simply 0; we can forget about this part of the expression. We now take account of the $\frac{1}{2}$ in front of the bracket, and find that our expression boils down to $a^2 + \frac{1}{2}(X_1{}^2 + X_2{}^2)$. Now $\frac{1}{2}(X_1{}^2 + X_2{}^2)$ is the thing we are interested in calculating – the average value of X^2. From page 311 on, we have denoted it by $\sigma_x{}^2$. Our expression accordingly may be written as $a^2 + \sigma_x{}^2$. This is a very simple result and may be summed up as follows; if instead of working out the average value of X^2, which would give $\sigma_x{}^2$, we work out the average value of x^2, we always get an answer that is too large, and in fact is too large by the amount a^2.

A numerical example may help to bring out the meaning of this. Suppose $x_1 = 9$ and $x_2 = 11$. The average of these two numbers is $a = 10$, so $X_1 = -1$ and $X_2 = +1$. Thus the direct way to find $\sigma_x{}^2$ is to find $\frac{1}{2}\{(-1)^2 + (+1)^2\}$ which is 1. In this particular instance, of course, this would be the easiest and most convenient way to calculate $\sigma_x{}^2$. If however someone for some reason found the average value of x^2 this would be $\frac{1}{2}(9^2 + 11^2) = \frac{1}{2}(81 + 121) = 101$, which exceeds the correct answer, 1, by exactly 100, which indeed is a^2. One could arrive at $\sigma_x{}^2$, though it would not be an economical method in this particular case, by getting 101 and then making the appropriate correction, that is, by subtracting a^2. This would give $101 - 100$, which is 1.

With only two numbers, the average cannot involve any fraction worse than $\frac{1}{2}$, so that it is impossible to give an example for which the result just found gives any real arithmetical advantage. Perhaps $x_1 = 0$, $x_2 = 1$ is as good as any. Then the average value of x^2 is $\frac{1}{2}$. As the average a is $\frac{1}{2}$, to get $\sigma_x{}^2$ we must subtract a^2, which is $\frac{1}{4}$. This gives $\frac{1}{2} - \frac{1}{4} = \frac{1}{4}$ for $\sigma_x{}^2$. It is easily verified by using the more direct method that this answer is correct.

We have now dealt with the first part of our problem; we have shown how the average of x^2 compares with the average of X^2. An exactly similar result holds of course for the average of y^2. Since the average of the values of y is b, the average of y^2 will be $b^2 + \sigma_y{}^2$.

It remains only to deal with the third part of the problem, to compare the averages of xy and XY. The average of xy is $\frac{1}{2}(x_1 y_1 + x_2 y_2)$. We have $x_1 = a + X_1$ and $y_1 = b + Y_1$. Multiplying these together we can find $x_1 y_1$. Similarly $x_2 = a + X_2$ and $y_2 = b + Y_2$, and we have to multiply these together. Altogether we get eight terms, and we thus find we have to sort out the following expression for the average of xy:

$$\frac{1}{2}(ab + bX_1 + aY_1 + X_1 Y_1 + ab + bX_2 + aY_2 + X_2 Y_2).$$

It helps you to look at this if you realize that the first four terms come from the first multiplication and the last four terms from the second multiplication.

Now again we have the question of the best way of grouping these terms. It is best to fix attention on the letters a and b. Inside the bracket we see $2ab$, which is simple enough. The terms which contain neither a not b are $X_1 Y_1 + X_2 Y_2$ and we cannot simplify these in any way; we can only copy them down. Now four terms remain for consideration; two of them contain a and two contain b. The two terms that contain b are $bX_1 + bX_2$, which is the same as $b(X_1 + X_2)$. Just as in the work a little earlier, we note that $X_1 + X_2 = 0$, so these terms give $b \cdot 0$ which is 0. The two terms containing a behave similarly. They are equal to $a(Y_1 + Y_2)$. As we first observed in question (v) of the exercises and later showed how to prove, the sum of the Y values is 0. So here again we get 0.

Accordingly all that survives of our long expression is $\frac{1}{2}(2ab + X_1Y_1 + X_2Y_2)$, which is the same as $ab + \frac{1}{2}(X_1Y_1 + X_2Y_2)$, and this we can express in words as ab added to the average of XY. This then is the average of xy. Accordingly, if we have computed the average of xy, and wish to find the average of XY, all we have to do is to subtract ab.

Now all of this work, of course, has dealt with the particular case in which only two points were involved. Fortunately, when we go on to the case of three points, no new principle is involved. Instead of frequently seeing the fraction $\frac{1}{2}$ we shall see the fraction $\frac{1}{3}$. The expressions will be rather longer. That giving the average of xy will contain twelve terms instead of eight; the first four terms arise from multiplying out x_1y_1, the next four from x_2y_2 and the last four from x_3y_3. The method of grouping we used a few lines earlier will work equally well and in fact will lead to the identical conclusion; the average of xy is the result of adding ab to the average of XY. The results for the averages of x^2 and y^2 also remain true, and can be proved by algebra that involves a little more writing but no new idea. And in fact the same is true if we have four points, or five or any number.

The effect of the three results obtained above is to allow us to work from points other than the average point G. At the end we must come back to G since, for instance, the slope of the line is the average of XY divided by the average of X^2, and X and Y must be measured from G. The results above allow us to work with coordinates x, y measured from some other point and at the end to make the corrections needed. The origin from which x and y are measured need not be that to which the original data refer. For example, our comparison of the climates of Aberdeen and Montreal was based on the following figures:

| Aberdeen | 42 | 40 | 43 | 50 | 55 | 60 | 63 | 63 | 59 | 53 | 47 | 43 |
| Montreal | 26 | 25 | 33 | 42 | 56 | 66 | 73 | 62 | 65 | 54 | 42 | 31 |

The average of the Aberdeen figures is 51·5, of the Montreal figures 47·92. We can get close to G without getting away from whole numbers if we use 51 and 48. The following table shows the excess of the Aberdeen temperature over 51 as x and the excess of the Montreal temperature over 48 as y:

x	-9	-11	-8	-1	4	9	12	12	8	2	-4	-8
y	-22	-23	-15	-6	8	18	25	14	17	6	-6	-17

With this system of coordinates, we find a, the average value of x, is 0.5 and b, the average value of y, is -0.08. The average value of x^2 is then 66.67. To get the average value of X^2 we have to subtract a^2, that is $(0.5)^2$. This gives 66.42, which accordingly is σ_x^2. Hence $\sigma_x = 8.15$. The average value of y^2 is 261.08. From this we have to subtract b^2, which is 0.0064, so small that it hardly affects the answer. We thus find $\sigma_y^2 = 261.1$ approximately and $\sigma_y = 16.16$. The average value of xy is 128.92. To get the average value of XY we must subtract ab. As ab is -0.04, this means we must add 0.04, and we obtain 128.96.

VARIATIONS BETWEEN AUTHORS

One thing a reader should be prepared for is that various authors may state a formula or a principle in slightly different ways. For instance, nearly all the discussion in this chapter has been in terms of *averages*, – the average value of X^2, of XY, of x^2, and so on. Now to find an average, you add certain quantities together and then divide by their number, N. Accordingly some authors, instead of speaking of the a, the average value of x, may prefer to express this as the sum of the x values, divided by N. This may appear in symbolic form as $(\Sigma x)/N$. Similarly, the average value of X^2 may appear as $(\Sigma X^2)/N$. Thus formulas expressed in terms of sums rather than averages may contain the symbol N in one or more places. I say 'may' and not 'will' because on occasion cancelling may occur. For instance (as you may verify) the slope of the line may equally well be expressed as the average of XY divided by the average of X^2 or as the sum of XY divided by the sum of X^2.

Coping with such variations of practice is a good example of how a reader of scientific works is continually compelled to use a little simple algebra. If a book is dealing with correlation it is likely to be doing something like what we are doing in this chapter. But it may use different symbols or work in terms of slightly different quantities. The reader will need to work out any unfamiliar quantity in terms of quantities he is accustomed to use. After substituting the expressions found, the unfamiliar formula ought to prove to be a familiar formula in disguise.

In this chapter we have followed Galton's thinking and arrived at correlation in terms of 'the dotted line' or *line of regression,* as it is more technically known. Quite apart from historical interest or any insight gained by this approach, there is some practical value in doing this. We do sometimes want to find the line that fits data as closely as possible. However there are occasions when we simply want to know r, the coefficient of correlation, and are not bothered about the slope of the line. With such occasions in mind we wish to find a formula that gives r directly, and we can do this by putting into algebraic form the procedure we have been following in this chapter. We will need a symbol, say K, for the average value of XY. As we saw on page 315, the slope of the line is K divided by the average of X^2. As the average of X^2 is denoted by σ_x^2, the slope of the line is thus K/σ_x^2. The change of vertical scale then causes this fraction to be multiplied by σ_x/σ_y. This would give $(K\sigma_x)/(\sigma_x^2\sigma_y)$, which we can simplify, dividing above and below by σ_x. The final result is thus that $r = K/(\sigma_x\sigma_y)$.

A small point of conventional textbook algebra arises in calculating r from the formula we have just found. In question 3, in the exercises earlier in this chapter, one finds $\sigma_x = 3\cdot16$, $\sigma_y = 5\cdot92$, and we can if we like just multiply these together to get the denominator in the formula for r. However, another procedure is possible. As $\sigma_x = \sqrt{10}$ and $\sigma_y = \sqrt{35}$, $\sigma_x\sigma_y = \sqrt{10}.\sqrt{35}$ and, as we show on page 329, this is the same as $\sqrt{350}$, which is $18\cdot71$. The advantage of this last procedure is that we only look one number up in the square root tables instead of two, and also that the multiplication involves much easier numbers. Now of course one could survive without knowing this device. The case for knowing it, I suppose, rests on a combination of the following considerations: it may on occasion allow you to make a quick calculation when your electronic computer or slide rule is not handy; it provides an alternative method which can be used to check a computation, and the more automatic devices you use, the greater is the need to check that nothing has gone wrong with them; the writer of a textbook or a scientific paper may use it, and you will be puzzled if you do not know what he is doing. One might add that the principle involved is not a very complicated one nor should it require a great amount of time to learn.

A NOTE ON CLUSTERS

Earlier we referred to clusters being egg-shaped and drew an ellipse to represent the egg in question. The difficulty may have occurred to some readers; how do we decide where the boundary of the cluster is? Even if we go very far away from the centre, we may still find on our diagram an occasional dot which represents some very infrequent event. Must the ellipse be big enough to include such rare events? If not, how do we decide at what stage the dots become rare enough to be neglected?

Fortunately, we do not have to make such a decision, for in fact we have to deal not with a single ellipse but with a whole sequence of ellipses. Figure 129 is drawn on the same principle as some maps that show density of population. The darkest region

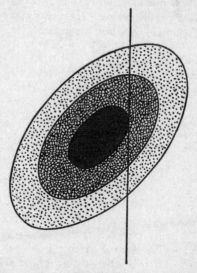

Figure 129

indicates very dense population; the shaded, the dotted, and the unmarked regions show increasingly thinly populated areas. The population we are considering here is, of course, that of the dots on the diagram. The contours for population density are ellipses. All the ellipses have similar shapes; that is to say, any ellipse can be obtained from any other simply by changing the scale of the drawing. In our earlier argument, we supposed a value fixed for x, and considered the average value of y for the points with this x coordinate. (See page 305.) In Figure 129, the vertical line corresponds to the choice of such an x. As we go upwards along this line, we start in the unmarked region of very thin population, and then find population density increasing as we pass through the dotted and shaded zones. We just touch the darkest zone, and then population density begins to decrease as we ascend further through the shaded, the dotted, and finally the unmarked zones. It can be shown that the average value of y for dots on this line is at the point where the line touches the black ellipse. If we picked any ellipse, found where the line crossed it, and took the midpoint, we should arrive at the point of contact with the black ellipse. Thus the ellipse shown in Figure 124 could be any one of the ellipses in our population map. The construction used in connexion with Figure 124 was therefore justified.

CHAPTER FOURTEEN

On Square Roots, Irrationals, and All That

CHAPTER 6 of *Vision* dealt with one of the most widely known results in mathematics, Pythagoras' Theorem, according to which the sides of a right-angled triangle are connected by the equation $x^2 + y^2 = z^2$. This formula is very easy to use when everything works out in whole numbers. For instance, if $x = 12$ and $y = 5$, we have $z^2 = 144 + 25 = 169$, and as 169 is 13^2 we know immediately that $z = 13$. However, it is long odds against things working out as easily as this. Usually $x^2 + y^2$ will come out as a number which is not the square of some other whole number. For instance, as we saw on page 138 of *Vision*, if $x = 6$ and $y = 4$, we get $x^2 + y^2 = 52$, which falls between 49 and 64, the squares of 7 and 8. Fortunately, as was mentioned there, tables of square roots are available which show that $z = 7 \cdot 2111$ is (to four places of decimals) the number that makes $z^2 = 52$. We have to find square roots so often that a special sign has been devised to save writing. This sign is $\sqrt{}$. If we write $7 \cdot 2111 = \sqrt{52}$, this conveys very briefly what we have just said at length – that if you are looking for a number, which when squared gives 52, then $7 \cdot 2111$ is the number you want. When a learner of mathematics meets a new sign, it usually takes him a little while to get used to it, so books on arithmetic and algebra usually give a few routine exercises, which are not intended to test powers of calculation, but merely to make sure that the reader understands the meaning of this sign. Such a question might be, 'What is $\sqrt{25}$?' The answer is 5, since $5^2 = 25$. Most readers of this book will probably be familiar with the sign $\sqrt{}$, but it will not need much space to give a few such questions and provide answers at the back of the book, so this is done in case any readers are not familiar with the sign, or have met it and had difficulties with it.

Exercise

Find the following numbers: (i) $\sqrt{9}$, (ii) $\sqrt{100}$, (iii) $\sqrt{16}$, (iv) $\sqrt{36}$, (v) $\sqrt{49}$, (vi) $\sqrt{1}$, (vii) $\sqrt{\frac{1}{4}}$.

If we have the equation $z^2 = 25$, there are two numbers that meet this requirement, namely $+5$ and -5. It is usually agreed that when we write $\sqrt{25}$ we mean the positive number $+5$, not the negative number -5. For someone who has not yet heard about negative numbers the equations $z^2 = 25$ and $z = \sqrt{25}$ both say exactly the same thing, namely that z is a number which when squared gives 25.

USING THE TABLES

When you look at a table of square roots, you will usually find that the table goes from 1 to 100 and then stops. The question naturally arises – what do you do if you want to find the square root of a number, say, between 1,000 and 10,000? Fortunately, it is not necessary to have tables which go on and on indefinitely. We can find the square root of any number with the help of tables that go only from 1 to 100.

The reason for this can be seen from an example. Consider 4^2 and 40^2. We have $4 \times 4 = 16$, while $40 \times 40 = 1,600$. We notice the figures 16 in both, but the square of 40 is 100 times as large as the square of 4. A general result is involved here. If a is any number, the square of $10a$ is 100 times as large as the square of a. The truth of this follows immediately when we work out these squares. For $(10a)^2 = 10a \cdot 10a = 100a^2$, which is indeed 100 times a^2.

Suppose we want to know $\sqrt{5,200}$. We have already observed that $\sqrt{52} = 7 \cdot 2111$, which is the same as saying $(7 \cdot 2111)^2 = 52$. The result in the last paragraph tells us how to write down the square of $72 \cdot 111$, the number 10 times as large as $7 \cdot 2111$. Its square will be 100 times 52. So $(72 \cdot 111)^2 = 5,200$, and we have found the number which, when squared, gives 5,200.

Now of course we do not always have to deal with nicely rounded numbers such as 5,200. But the principle is exactly the same. If we want to know $\sqrt{5,273}$, we consider the fact that 5,273 is 100 times $52 \cdot 73$, so $\sqrt{5,273}$ will be 10 times $\sqrt{52 \cdot 73}$, and this we can find with the help of our tables.

It is to be noticed that any question about square roots can be changed into a question about squares. This is useful, as squaring a number is a fairly simple and familiar idea, while the square root is not something that people can visualize as easily.

On Square Roots, Irrationals, and All That

For instance, many learners find no trouble at all with the question. 'What is the square of ab?' They readily work out $(ab)^2 = ab.ab = a^2b^2$ by a routine similar to that described on page 209 of *Vision*. But the corresponding result, obtained by turning this work round and stating it in terms of square roots rather than squares, is not found so easily. We will obtain this result first of all by formal algebra, and then discuss it a little in terms of arithmetic. The result $(ab)^2 = a^2b^2$ involves a^2 and b^2. Suppose $a^2 = p$ and $b^2 = q$. Here are two statements about squares. We can turn them round and make them statements about square roots. If we were starting with p and q rather than a and b, we would probably think of these equations as stating $a = \sqrt{p}$ and $b = \sqrt{q}$. If we substitute p for a^2 and q for b^2 in our equation $(ab)^2 = a^2b^2$ it becomes $(ab)^2 = pq$. Here is another statement about a square, so we turn it inside-out and make it a statement about square roots, namely $ab = \sqrt{(pq)}$. In this we can substitute $\sqrt{p}$ for a and $\sqrt{q}$ for b so we get $(\sqrt{p}).(\sqrt{q}) = \sqrt{(pq)}$. In the experience of many teachers exercises involving an understanding of this result produce far more mistakes than exercises on squaring ab. Both results are saying the same thing, but the second seems to say it much less plainly than the first. The second result is important for all work involving square roots, and, as we saw in Chapter 10 and elsewhere, square roots are extremely liable to turn up in mathematical calculations. This result then is something that should not be rushed. Indeed, in some ways this result is a test for any system of mathematical education. To work with it successfully, a learner must be ready to do something that is really the key to all mathematical work; he must be willing to keep coming back to it, turning it over in his mind, considering examples of it and working with it until he is quite clear what it means. The pupil who has been hurried through a series of rules will almost invariably make glaring mistakes in this matter.

Our result is that $\sqrt{(pq)} = (\sqrt{p}).(\sqrt{q})$. We have already met an arithmetical example of this. If we put $p = 100$, then $\sqrt{p} = 10$ and our equation becomes $\sqrt{(100q)} = 10.(\sqrt{q})$. This was the theme of our earlier remarks on how to use the tables. The square root of $100q$ is only 10 times as big as the square root of q; $\sqrt{1,600}$ is not 100 times, but only 10 times as big as $\sqrt{16}$.

It sometimes happens that we want to find the square root of a number which arises in a form such as 36×49. Now of course we could multiply out, get 1,764, and then look for the square root of this. But this would be a waste of effort. By using our result with $p = 36$ and $q = 49$ we would see that $\sqrt{(36 \times 49)} = (\sqrt{36}) \times (\sqrt{49})$, which is simply 6×7 or 42. Here again the correctness of the result may be seen very easily by thinking in terms of squares rather than square roots. We start at the end and work backwards. The square of 42 is $(6 \times 7) \times (6 \times 7)$. As we saw in *Vision*, it does not matter in what order the multiplications are carried out. We shall get the same number if we work out $6 \times 6 \times 7 \times 7$. But this is just 36×49, the number we set out from.

When a number falls neatly into pairs of factors, as $6 \times 6 \times 7 \times 7$ does, finding the square root is simply a matter of splitting the product into two identical parts. Each pair of factors gives a factor to the first part and a factor to the second part. Thus $6 \times 6 \times 7 \times 7$ splits into 6×7 and 6×7, and is thus seen to be the square of 6×7.

The same idea can be applied to a less obvious case. Often, when you are reading some calculation in a book or scientific paper, you will meet an incidental step such as $\sqrt{50} = 5\sqrt{2}$. Now $50 = 5 \times 5 \times 2$, so the factor 5 is paired but we have a single factor 2. However, we can split this factor by introducing the sign $\sqrt{2}$, since $2 = (\sqrt{2})^2 = \sqrt{2} \times \sqrt{2}$. So $50 = 5 \times 5 \times \sqrt{2} \times \sqrt{2}$. This product can now be split into the identical parts, $5 \times \sqrt{2}$ and $5 \times \sqrt{2}$, each of which thus represents the square root of 50.

We can get in effect to the formula $\sqrt{(pq)} = \sqrt{p} \times \sqrt{q}$ by carrying this idea one stage further. If we are dealing with $\sqrt{6}$, since $6 = 2 \times 3$ we do not have any pair of factors. But we can achieve an effect of pairing, by using $3 = \sqrt{3} \times \sqrt{3}$ and $2 = \sqrt{2} \times \sqrt{2}$. Thus $6 = \sqrt{3} \times \sqrt{3} \times \sqrt{2} \times \sqrt{2}$, which splits into $\sqrt{3} \times \sqrt{2}$ and $\sqrt{3} \times \sqrt{2}$. So $\sqrt{6} = \sqrt{3} \times \sqrt{2}$. Needless to say, we do not use this result as a way of working out $\sqrt{6}$. As tables give $\sqrt{2} = 1 \cdot 414$ and $\sqrt{3} = 1 \cdot 732$, it is much easier to look up $\sqrt{6}$ in a table than to do the multiplication $1 \cdot 414 \times 1 \cdot 732$. It might perhaps be used in the opposite direction; if some calculation led to the answer $\sqrt{2} \times \sqrt{3}$, we would then simplify this to $\sqrt{6}$. It would also be useful if some answer came to be $\sqrt{6} \div \sqrt{3}$, which we would then see to be simply $\sqrt{2}$. Or, most likely, it would turn up as an incidental detail in some long calculation or theoretical argument.

In our next section we shall be particularly concerned with the square root of a fraction, so we now consider this question. We have just been considering a formula for $\sqrt{(pq)}$, in which p and q are multiplied. We have now to study $\sqrt{(p/q)}$ in which p is divided by q. This result, in exactly the same way, can be obtained by formal algebra and illustrated by examples in arithmetic. The confidence of a learner is increased if he can derive the formula for himself, and accordingly we set this question out as an exercise.

Exercise

All questions of square roots can be turned round and considered as questions of squaring. Now for any fraction a/b we have $(a/b)^2 = a^2/b^2$. Let $p = a^2$ and $q = b^2$. Our task is twofold; from the above equation we have to obtain an equation involving p and q only, with no mention of a and b; we also have to get an equation in which the operation of squaring is not mentioned, but only the finding of square root. Thus we must reach an equation in which $\sqrt{}$ can appear as often as it likes, but we must not see anywhere $(\ )^2$. Some readers may find the following outline of strategy helpful.

(i) Can you express the equation $(a/b)^2 = a^2/b^2$ in the form '... is the square root of ...'?

(ii) In the statement you find in (i), can you get rid of all mention of a and b, and get a statement purely in terms of p and q, where $p = a^2$ and $q = b^2$?

There are some things to notice about squaring a fraction which will be important in our next section. Consider, for example, the square of 10/21. As $10 = 2 \times 5$ and $21 = 3 \times 7$ we may write

$$\left(\frac{10}{21}\right)^2 = \frac{10}{21} \times \frac{10}{21} = \frac{10 \times 10}{21 \times 21} = \frac{2 \times 5 \times 2 \times 5}{3 \times 7 \times 3 \times 7}$$

We could if we liked write $(10/21)^2$ as 100/441 or as $10^2/21^2$.

First notice that the numbers 100 and 441 which occur as the numerator and denominator of $(10/21)^2$ are 10^2 and 21^2, both of them perfect squares (that is, squares of a whole number, not like 2 which is the square of $1 \cdot 414 \ldots$.) This must always happen when we square any fraction. Someone might object, 'I agree that, when you multiply the fraction a/b by itself, you get a^2/b^2, so you have the perfect square a^2 on top and the perfect square b^2 down below, but then there might be some cancelling so we would get smaller numbers above and below, and these might not be squares'. It was to meet this objection that we showed the factors of 10 and 21 at the beginning of this discussion. 10 has the prime factors 2 and 5 while 21 has 3 and 7. There is no common factor here, of course; if there were, we would remove it before we began the work of squaring the fraction. When we square, we get $2 \times 5 \times 2 \times 5$ above; here we see the same factors, 2 and 5, as we had in 10, but each appears twice; similarly, down below, we see the factors 3 and 7, each twice. Since 3 and 7 have nothing in common with 2 and 5, when each is repeated twice there will still be nothing in common. Accordingly, there cannot be any cancelling occurring after we have squared 10/21. Exactly the same argument applies to the squaring of any fraction, provided of course that we begin with the fraction in its lowest terms (that is, so written that no cancelling is possible). Accordingly we have our conclusion; if a/b is any fraction, where a and b are whole numbers and no cancelling is possible, the square of a/b will have perfect squares, a^2 and b^2, above and below.

THE UNMENTIONABLE OR THE UNNAMEABLE?

The conclusion at the end of the last section does not sound very remarkable or surprising. Yet it led to one of the great crises in mathematical thought.

Suppose you put the following question to any ordinary person. When we are measuring lengths, we sometimes get whole numbers, such as 7 inches, or we may get fractions, such as $2\frac{3}{4}$ inches. If we are measuring accurately for engineering purposes, we may need to say that a rod is 267 thousands of an inch long. Suppose, then, that we are familiar with all the fractions there are; will this be sufficient to measure any length whatever? The natural answer would be 'Yes'. By means of fractions we can make extremely fine distinctions. We can name lengths smaller than any-one working with the nucleus of an atom can measure. It would seem that fractions are perfectly adequate to express any measurement we may ever need.

Now in practical terms this is correct. In any actual measurement, there is always some element of error, so that the specification of a length beyond a certain degree of accuracy becomes physically meaningless. Fractions are perfectly adequate for all practical purposes. However in geometry there are certain lengths for which we have theoretical formulas. It was discovered by the ancient Greeks that the lengths of some lines could not be specified by means of fractions. An example of such a line would be the diagonal of a square with side of length 1 inch. (Figure 130.)

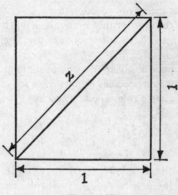

Figure 130

The length of the diagonal is easily found from Pythagoras' Theorem. We have $z^2 = x^2 + y^2$ with $x = 1$ and $y = 1$, so $z^2 = 2$. The length of the diagonal is given by a number whose square is 2. It would then be natural to ask, 'What number is that?' And here we run into difficulties. We can get near it. The square of 7/5 is 49/25, which is just under 2. So z is a little more than 7/5. But what is z *exactly*? We cannot answer this by giving any fraction whatever, for as we have seen, when you square any fraction, you get perfect squares above and below. Now 2 may be written as the fraction 2/1. We have the perfect square, 1, below, but on top we have 2, which is not the square of any whole number. *It is therefore impossible that any fraction whatever should give 2 exactly.* We cannot express $\sqrt{2}$ as a fraction, in the sense in which fractions are understood in elementary arithmetic, i.e. one whole number divided by another.

This discovery must have surprised the ancient Greeks and is said to have disturbed them. According to one story, the Pythagoreans forbade their members to disclose this result to the public and referred to $\sqrt{2}$ and similar numbers as 'the unmentionable'. I am not familiar with the research on Greek mathematics and would not like to make any dogmatic statement about this story. However I have noticed that stories about the history of mathematics do seem to improve with the telling, and the possibility has occurred to me that the Pythagoreans may have been using a Greek word that meant not so much 'the number that may not be mentioned' as 'the number that cannot be named'. Whatever the history may be, the latter interpretation would give a good description of the number $\sqrt{2}$.

While $\sqrt{2}$ cannot be given by a single fraction, there is a sequence of fractions that specifies it quite accurately. This sequence is formed by a simple rule. The sequence is

$$\frac{1}{1}, \frac{3}{2}, \frac{7}{5}, \frac{17}{12}, \frac{41}{29}, \frac{99}{70}, \ldots$$

The numerators are 1, 3, 7, 17, 41, 99, ... To get these, you have to write down the first two, 1, 3. To get the next number you double 3 and add 1. This gives 7. Then to get 17, you double 7 and add 3. That is, each number is twice the number before it, together with the number that came before that. The denominators are formed by the same rule. Thus, for example, 70 is found by doubling 29 and adding 12.

These numbers are alternately too small and too large for $\sqrt{2}$. Thus $(7/5)^2$ is 49/25, which is less than 2 by 1/25. Then $(17/12)^2$ is 289/144, which exceeds 2 by 1/144. The next square, $(41/29)^2$ is too small by 1/841 while the following square $(99/70)^2$ exceeds 2 by only 1/4,900.

However far we may go, we always find the square is a little too big or a little too small. No fraction in this sequence gives the exact value of $\sqrt{2}$. But the fractions close in on $\sqrt{2}$, some from the right and some from the left, as shown in Figure 131. From the first pair of fractions, we see only that $\sqrt{2}$ must be somewhere between 1 and 1·5 (which is 3/2). As soon as we take account of 7/5, the left-hand fence moves in to 1·4. At each step the fences move closer together, like the walls of a medieval torture chamber. They never actually meet, but by going sufficiently far down the list of fractions, you can make the distance between fences as small as you like. There is only one number that is sure to be between them, however much of the list we may consider. That number is $\sqrt{2}$. Any other number, however close it may be to $\sqrt{2}$, will find itself outside the fenced portion sooner or later.

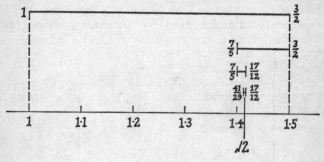

Figure 131

Thus $\sqrt{2}$ is a number that cannot be specified by a single fraction, but only by giving an unending series of fractions which are better and better approximations to it. There are many other numbers of which the same can be said. They are referred to as *irrational numbers*. As language is used today, this tends to suggest that they are unreasonable, emotional, or insane numbers. This, of course, is not what is intended by this term. The word *ratio* comes from Latin in which it had many meanings. One of its principal meanings was any kind of calculation. In arithmetic, it seems to have settled down to one particular kind of calculation, finding the proportion of one number to another, as in 'the ratio of 7 to 5'. This is the meaning relevant to our present purpose; an irrational number is one that cannot be specified as the ratio of one whole number to another.

There are many such numbers. Any square root or cube root that does not 'come out neatly', such as $\sqrt{7}$ or $\sqrt[3]{5}$, is irrational. So is the number π, which we meet early in life in connexion with the circle. The area inside a circle is given by πr^2. For a circle of unit radius, $r = 1$, and the area is simply π. This is a convenient way of visualizing what π means.

Suppose we have drawn a circle of unit radius. How shall we go about measuring its area? It is not too clear what we mean by the area of a circle. We think of the area of a region as the number of square feet or square inches or square centimetres needed to cover that region. Now a circle is a remarkably uncomfortable thing to cover with squares, or even with rectangles. I mention rectangles, because the well-known formula $A = hb$ (often memorized as 'area equals height times base') gives a definite meaning to the area of a rectangle.

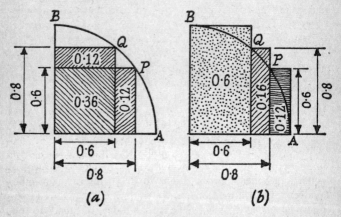

Figure 132

It is obviously impossible to break a circle up into a number of rectangles, but we can do what we did with $\sqrt{2}$. We can produce a sequence of numbers that close in from above and below.

How to go about this is shown in Figure 132. Both parts of the figure show a quarter of a circle of unit radius with centre at the origin, O. P and Q are two points on this circle. The choice of these points is suggested by our knowing that there is a right-angled triangle with sides 3, 4, 5. The point Q has $x = \frac{3}{5}$ and $y = \frac{4}{5}$. It can be checked by Pythagoras' Theorem that the distance OQ is 1, so Q is on the unit circle. The point P has $x = \frac{4}{5}$ and $y = \frac{3}{5}$. It will be agreed that the shaded area in Figure 132(a) lies inside the quarter cirle. If the quarter circle were covered with material, this shaded region could be cut out from it, so presumably has less area than the quarter circle. The area of the shaded region is 0·60, while that of the quarter circle is $\pi/4$. So π must be more than 4 times 0·60, which is 2·40. Similarly, by looking at Figure 132(b), we see the quarter circle covered by a shaded region, consisting of three rectangles with total area 0·88. So $\pi/4$ must be less than 0·88, which means π less than 3·52. So π lies between 2·40 and 3·52. This of course everyone knows anyhow. The point is that by carrying this method further, putting more and more rectangles inside the circle in (a) and cutting down the waste outside the circle in (b) we could get as near as we like to π, but we should never quite get there.

Many of the numbers that occur in more advanced mathematics are like π. You cannot write them down as a fraction, but you can give fractions that pin them down more and more closely.

It should be mentioned that the considerations above do not *prove* that π is irrational. If you take the region enclosed by *any* curve, it is impossible to divide that region into rectangles, and we have to give an unending sequence of overestimates and underestimates *in order to show what we mean* by the area inside a curve. But it can happen that the number caught between these overestimates and underestimates, as they close in, turns out to be a rational number, an ordinary fraction. In fact to prove that this does not happen with π is quite a difficult job. I can only assure you that mathematicians are agreed that π is indeed an irrational number.

IRRATIONALS AND THE SCHOOL SYLLABUS

The discussion above shows that producing a number to specify a length or an area is a much more complicated business than one would at first suppose. There are in fact three possibilities; the number may be a whole number, such as 5; it may involve fractions, like $7\frac{1}{2}$; it may be an irrational number, like $\sqrt{2}$ or π, that can be approached, but never reached, by an unending series of fractions.

As we saw in *Vision*, there are great advantages both in algebra and in its application to geometry, in using the idea of negative numbers. If we allow whole numbers, with positive or negative signs, we are working with the collection of numbers . . . -3, -2, -1, 0, 1, 2, 3, . . ., known as the *integers* (from *integer*, Latin for *whole*.) If we accept these and also fractions with positive or negative signs, we are working with the *rationals*. Finally, if we allow irrationals with either sign, such as $\sqrt{2}$, π, $-\sqrt{2}$, $-\pi$, we are working with the *real numbers*. The real numbers are often visualized as being spread out along a line, so that we may say -5 is to the left of -2 or that π is to the right of 3.

Much advanced mathematics, in particular calculus, involves real numbers. From time to time difficulties have been encountered which have made mathematicians want to be clearer in their own minds just what the real numbers are, and what properties of them may be used in constructing a proof. In this way an elaborate and extensive theory has been built up, to which university students are exposed, in some measure, in courses on calculus, analysis, or real variables. In these courses an attempt is usually made to start with certain clear statements, and to deduce everything else from these in a strictly logical manner. In such a course one moves steadily forward like a vehicle travelling on a clear, straight road. Unfortunately the development of mathematics, both in the history of its discovery and in the education of an individual, does not move like this, but rather like a tide which advances, retreats, works over the same ground again, and finally advances a little further. For an example of this we need not go outside the topic of the present chapter. To the best of my knowledge, before the discovery of Pythagoras' Theorem, nobody had any reason to study $\sqrt{2}$, and most certainly nobody had any inkling that there were irrational numbers. Accordingly the early proofs of Pythagoras' Theorem must have ignored the logical difficulties that can arise from some of the sides of the triangle having irrational lengths. Just where the difficulty was slurred over would depend on the particular approach used. In *Vision* a way of arriving at Pythagoras' Theorem was used that depended on the algebraic result $(x + y)^2 = x^2 + 2xy + y^2$, but this result was based on arguments in which x and y stood for *whole numbers only*; we visualized xy in terms of x rows each containing y objects. Later in *Vision* (around page 323) we had diagrams for multiplication of fractions, which could perhaps have been developed to justify the algebraic result in the case where x and y are fractions. But nowhere in *Vision* did we try to justify that result for irrational x and y. And in fact, it would have been entirely out of place to do so, since *Vision* was planned for children of 9 or 10 years of age or for beginners in mathematics. Neither type of reader would have any idea that there was such a thing as an irrational number. The discussion of irrational numbers would be far too sophisticated for learners at that stage; they would find it lengthy, boring, incomprehensible, and pointless. Only after Pythagoras' Theorem has become familiar can the concept of irrational number arise in a natural way.*

Yet, right from the beginning, we apply Pythagoras' Theorem to cases involving irrational numbers. The right-angled triangles with whole number sides are exceptional: (3, 4, 5), (5, 12, 13), and so on. Any practical problem is likely to land us on irrational numbers, as it did already on page 141 of *Vision* with $y = \sqrt{8,001}$. It is perfectly feasible, as was done in that chapter, to use irrational numbers without ever mentioning them.

The early teaching of mathematics – which means not only the teaching of children but also the teaching of those who only intend to master a little of the subject – thus cannot be a rigorous, logical demonstration of the kind expected in a university course for mathematical specialists. Its object is not to provide proofs that meet the stringent requirements of a modern mathematician. Rather it is to make the learners familiar with mathematical situations, so that they can imagine these and think about them *in the way a mathematician thinks when he is discovering a new result and before he has been able to find a proof for it*. Such teaching is a delicate undertaking, for we are by no means indifferent to logic. We are not trying to establish in our pupils the habit of accepting statements without proof. Quite the reverse, we are

---

* In U.S.A. there have been attempts to introduce irrationals early by calling attention to non-recurring, non-terminating decimals. This is extremely artificial and the logical difficulties are at least as great.

trying to encourage them to argue with each other and to reason more and more closely about all ideas submitted to them. But this must be a development that takes place *in their minds*. It is not helped by drawing attention to subtle objections and difficulties, based on possibilities outside their experience and beyond the stage of their mathematical maturity.

A person can only think effectively at the level of philosophical analysis when he is aware of a real problem, doubt, or difficulty within his own mind. It was a weakness of many traditional courses in geometry that pupils wrote proofs more to please their teachers than because they felt the results needed proving. It is also a weakness of many university courses on analysis that they embark on an elaborate study of the real line without first having convinced the audience that you get into trouble if you do not make such a study. In fact the hardest thing in preparing a course on analysis is to assemble material that will make students feel that here is muddle that has got to be cleared up somehow.

These remarks on the relationship of advanced study to elementary teaching seem to be needed, because there has been in various countries a movement (now to some extent on the wane) to inject the university organization of material into elementary and secondary schools. Particularly on the American continent, there have been education authorities who imposed on the schools courses of study in which everything was organized around such topics as The Set of the Whole Numbers, The Set of the Integers, The Set of the Rationals, and so forth. At some stage of mathematical study it may be quite sound to define in turn the integers, the rationals, and the real numbers, and to prove that the same algebraic procedures apply to all of them. It is however quite inappropriate to take such a scheme as a skeleton around which to build elementary and secondary mathematical education. This is particularly harmful for a pupil with scientific or practical interests that call for only a limited amount of mathematics. With such a pupil, we continually have to demonstrate that the mathematics we are doing is interesting and relevant to his concerns. Now it would be extremely hard to find practical situations in which one needed to know that such things as irrational numbers existed. A passing reference may be useful. Pythagoras' Theorem has many obvious uses in practical work. So, for instance, a technical student might be warned that when he sees 1·414 in a table as being $\sqrt{2}$, he should not expect that squaring 1·414 will give him exactly 2, since 1·414 can be written as a fraction (it is $\frac{707}{500}$ in fact) and there is no fraction that gives $\sqrt{2}$ exactly.

There is a continual movement of mathematics down from the universities into the schools, and this will continue and is to be welcomed. It is also desirable that school teachers should keep an eye on university mathematics and science, in order to get new ideas and to prepare those pupils who are going on to more advanced work. But it is rarely satisfactory for schools simply to copy universities. Their tasks are fundamentally different. Ideas taken from advanced work have to be digested, rearranged, re-organized before they can be widely used in schools. This can only be done if there are teachers in the schools who are very competent mathematically and at the same time have a strong feeling for good teaching. Only in countries where a strong core of such teachers has been available to give effective leadership have reforms in mathematical education worked out satisfactorily.

Symbols

/   Used by printers so that fractions can be shown on a single line. Thus $\frac{3}{4}$ is replaced by 3/4. Has same meaning as $\div$. See page 94.

$>, <, \geqq, \leqq, \geqslant, \leqslant, \neq$. See page 118.

$\sqrt{}$ The square root of. $\sqrt{9}$ means a number which, when squared, gives 9. As $(3)^2 = (-3)^2 = 9$, both 3 and -3 appear to be candidates. A convention is usually adopted that only the positive solution is acceptable, so $\sqrt{9}$ indicates 3.

$\sqrt[3]{}$ The cube root of. $x = \sqrt[3]{a}$ means $x^3 = a$. For real numbers a and x, this equation has only one solution, so no convention about sign is required. For example, $\sqrt[3]{1,000} = 10$, $\sqrt[3]{(-125)} = -5$.

$\sqrt[n]{}$ The nth root of. $x = \sqrt[n]{a}$ means $x^n = a$. To avoid difficulties with convention, this symbol usually appears in contexts where negative numbers are excluded.

Σ Summation sign. See page 124.

∞ Infinity. Brief mention, page 124.

!   Factorial. See page 187.

$|x|$ The distance of x from 0. Known as *absolute value* or *modulus*. As 3 and -3 are both at distance 3 from 0, each has absolute value 3. In symbols, $|3| = 3$, $|-3| = 3$.

$f(x)$ There are considerable variations in the use of this symbol, depending both on the date at which a book was written and the branch of mathematics being discussed. On page 18 this symbol is discussed in relation to the idea of *function*. In algebra books it may often be found as indicating any polynomial in x. (Compare the symbol P(x) in the last pages of Chapter 7.) $f(7)$ would then indicate the result of substituting $x = 7$ in this polynomial.

Jargon

JARGON is a necessary evil when communication is made through the printed page. In schools, there should never be any substantial period devoted to the learning of jargon. Rather a teacher should casually use the words in the course of his regular exposition, so they become understood through their use. For a reader, the best procedure is to use this section only as a reference source; to look up a word in it only when some uncertainty about its meaning has arisen in the course of reading.

The *powers* of a number are obtained when a number is repeatedly multiplied by itself. Thus 2, 4, 8, 16, 32, ... are powers of 2. As $32 = 2 \times 2 \times 2 \times 2 \times 2 = 2^5$, we may refer to 32 as the 5th power of 2, or '2 to the 5th'. The earlier powers have special names dating from antiquity. Thus x^2 is usually called the *square* of x and x^3 the *cube* of x.

An expression such as $2x^4 - 7x^3 + 4x^2 + 8x - 11$ is called a *polynomial* in x. (Derivation essentially indicates *something with many parts*. The root *poly* for *many* is well known.) The parts $2x^4$, $-7x^3$, $4x^2$, $8x$, and -11 are known as the *terms*. The term in which x does not appear (here -11) is called the *constant term*. In all the other terms we see a power of x multiplied by a number which is called a *coefficient* (the derivation sounds rather slangy; *that which makes along with*). Thus 2 is the coefficient of x^4 since $2x^4$ is the first term of the polynomial; similarly -7 is the coefficient of x^3, since $-7x^3$ is seen in the polynomial. We may not be able to 'see' a coefficient which is 1 or 0. For instance, in $x^3 + 2x$, the coefficient of x^3 is 1, the coefficient of x^2 is 0 and the constant term is 0. We could, if we particularly wanted to, bring this out by writing $x^3 + 2x$ in the form $1.x^3 + 0.x^2 + 2.x + 0$.

The *degree* of a polynomial tells us what is the highest power occurring in it. In the paragraph above, the first example is a polynomial of degree 4, since x^4 occurs, but no higher power. The second example, $x^3 + 2x$, is of degree 3, since it contains x^3 but no higher power. Here again special names are used in certain cases. A polynomial of degree 3 is commonly called a *cubic* (it contains x cubed); a polynomial of degree 2 a *quadratic* (*quadra* is Latin for *a square*). $x^3 + 2x = 0$ is a *cubic equation*; $x^2 - 3x + 2 = 0$ is an example of a *quadratic equation*.

The number 4 that occurs in x^4 is spoken of as an *index* (plural, *indices*) or an *exponent*. In the seventeenth century it was discovered that meanings could be attached to fractional and negative indices;

343

thus it became possible to speak of x^{-1} or $x^{\frac{1}{2}}$ (see a sustained discussion in Chapter 3). However an expression is not regarded as being a polynomial if it contains fractional or negative powers.

$\dfrac{1}{x}$ is known as the *reciprocal* of x.

Answers to Exercises

Page 95

1. 5. Addition. $c = b + g$. 4. Subtraction. $g = c - b$.
2. 7. $x = t - y$.
3. £8. $s = p - r$. $p = s + r$. $r = p - s$.
4. $i = t + a$. $i = a + t$. $t = i - a$. $a = i - t$.
5. 70. $n = rx$. 4. $x = n/r$. 5. $r = n/x$.
6. $A = hb$. $b = A/h$. $h = A/b$.
7. (i) $C = V/R$; (ii) $R = V/C$.
8. (i) $v = p/m$; (ii) $m = p/v$.
9. 100. $t = an$. 4. $n = t/a$.
10. $a = p - b - c$ or $a = p - (b + c)$.

Page 110

1. $x = 5, y = 5$. 2. $x = 4, y = 4$. 3. $x = 11, y = 10$.
4. $x = 5, y = 4$. 5. $x = 11, y = 1$. 6. $x = 8, y = 2$.
7. $x = 7, y = 3$. 8. $x = 6, y = 4, z = 4$. 9. $x = 4, y = 6, z = 5$.
10. $x = 9, y = 1, z = 2$. 11. $x = \frac{1}{2}, y = \frac{1}{2}$. 12. $x = \frac{2}{3}, y = \frac{1}{3}$.
13. $x = \frac{3}{4}, y = \frac{1}{4}$. 14. $x = a/(a + 1), y = 1/(a + 1)$.
15. $x = ab/(a + 1), y = b/(a + 1)$.
16. $x = \frac{1}{2}b + \frac{1}{2}a, y = \frac{1}{2}b - \frac{1}{2}a$.
17. $x = a/(a + b), y = 1/(a + b)$.
18. $x = a/(ab + 1), y = 1/(ab + 1)$.

Page 116

1. $x = 4, y = 3, z = 2$.
2. $x = 1, y = 2, z = 1, t = 1$.
3. $x = 1, y = 3, z = 1, t = 3$.
4. $x = 3, y = 2, z = 1, t = 3$.
5. $x = 1, y = 1, z = 2$.
6. $x = 3, y = 2, z = 1, t = 3$.
7. $x = 2, y = 15, z = 12, t = 6$.
8. $x = 1, y = 1, z = 1, t = 1$.
9. $x = 1, y = 4, z = 1, t = 2, u = 2$.
10. $x = 4, y = 12, z = 10, t = 6, u = 1$.

The Search for Pattern

1. (i) L_1; (ii) L_3; (iii) D_2; (iv) D_1; (v) D_0; (vi) L_0.
2. (i) 18; (ii) 14; (iii) 4.
3. Subtraction; (ii) $4 = 18 - 14$; (iii) $L_3 = L_2 - D_2$.
4. (i) Yes; (ii) $L_2 = L_1 - D_1$.
5. (i) L_{n+1}; (ii) $L_{n+1} = L_n - D_n$.
6. (i) 15; (ii) $33 - 18 = 15$; (iii) $L_1 - L_2 = D_1$;
   (iv) $L_n - L_{n+1} = D_n$; (v) Certainly correct.
7. (i) $111 + 15 + 14 = 140$; (ii) $D_0 + D_1 + D_2$;
   (iii) $144 - 4 = 140$; (iv) $L_0 - L_3$; (v) $D_0 + D_1 + D_2 = L_0 - L_3$.
8. (i) $111 + 15 + 14 + 2 + 1 = 143$;
   (ii) $D_0 + D_1 + D_2 + D_3 + D_4$; (iii) $144 - 1 = 143$;
   (iv) $L_0 - L_5$; (v) $D_0 + D_1 + D_2 + D_3 + D_4 = L_0 - L_5$.

1. 1,000.   2. 8.   3. 25.   4. 111.   5. 23.   6. 1.   7. 0.   8. 0.

1. Answers can be found from equation i of this chapter, by substituting appropriate values of p, q, r, s.
   (i) $p = 0$, $q = 0$, $r = 6$, $s = 6$. Simplifies to $x^3 - x$.
   (ii) $p = 0$, $q = 1$, $r = 2$, $s = 1$. Simplifies to $(x^3 + 3x^2 + 2x)/6$
   or $(x^3/6) + \frac{1}{2}x^2 + \frac{1}{3}x$, whichever you prefer.
   (iii) $p = 0$, $q = -1$, $r = 6$, $s = 12$; $2x^3 - 3x^2$.
   (iv) $p = 0$, $q = 0$, $r = 0$, $s = 1$. Answer $x(x - 1)(x - 2)/6$ or
   $(x^3 - 3x^2 + 2x)/6$, or equivalent.
2. (i) y table 0, 1, 5, 14, 30, 55 gives $p = 0$, $q = 1$, $r = 3$, $s = 2$.
   $y = \frac{1}{3}x^3 + \frac{1}{2}x^2 + (x/6)$.
   (ii) y table 0, 1, 9, 36, 100, 225, 441.
   $p = 0$, $q = 1$, $r = 7$, $s = 12$, $t = 6$. (See page 188.)
   $y = \frac{1}{4}x^4 + \frac{1}{2}x^3 + \frac{1}{4}x^2$. Answer can also be written as
   $\frac{1}{4}x^2(x + 1)^2$ or $[\frac{1}{2}x(x + 1)]^2$.
3. 0, 0, 2, 6, 12, 20; $x^2 - x$.
4. Table coincides with 1(i), the answer to which, $x^3 - x$, could be
   put in the form $(x - 1)x(x + 1)$.
5. 0, 6, 24, 60, 120, 210. These are 6 times the numbers in the 1(ii)
   table. The answer to 1(ii) could be put in the form $x(x + 1)(x + 2)/6$.
6. 0, 0, 1, 3, 6, 10; $y = \frac{1}{2}x^2 - \frac{1}{2}x$. The answers are just half those
   for question 3. We need to print two tickets, A to B and B to A for
   the railway, but for the players A against B is the same game as B
   against A.

Answers to Exercises

Page 216

1. 2, 4, 8, 16, 32, 64, if we include the signal in which all positions remain flat.
2. 3 possibilities for 1 game, 9 for 2 games, 27 for 3; 3^n for n games. $3^7 = 2,187$. 1 chance in 2,187.

Page 218

1. (i) 5; (ii) 4; (iii) 3; (iv) $5 \times 4 \times 3$.
2. (i) $6 \times 5 \times 4$; (ii) $7 \times 6 \times 5$; (iii) $n(n-1)(n-2)$.
3. (i) $6 \times 5 \times 4 \times 3$; (ii) $7 \times 6 \times 5 \times 4$; (iii) $n(n-1)(n-2)(n-3)$.

Page 233

1. (i) $x - 1)^2 + (y - 1)^2$.
   (ii) $(x - 1)^2 + (y - 1)^2 = 1$, or $x^2 + y^2 - 2x - 2y + 1 = 0$.
   (iii) $(1, 0)$, $(2, 1)$, $(1, 2)$, $(0, 1)$.
   (iv) Yes. Yes.
2. (i) $(x - 2)^2 + (y - 1)^2$.
   (ii) $(x - 2)^2 + (y - 1)^2 = 9$, or $x^2 + y^2 - 4x - 2y - 4 = 0$.
   (iii) $(5, 1)$, $(2, 4)$, $(2, -2)$, $(-1, 1)$.
   (iv) $\sqrt{18}$, or $3\sqrt{2}$.
3. The circle is $(x - 5)^2 + (y - 5)^2 = 25$ or $x^2 + y^2 - 10x - 10y + 25 = 0$. For diagram and list of points see Figure 172 on page 281 of *Vision*.
4. (i) $(x - a)^2 + (y - b)^2$.
   (ii) $(x - a)^2 + (y - b)^2 = r^2$.
   (iii) Circle, centre (a, b) and radius r.
5. (i) $x^2 + y^2$.
   (ii) $x^2 + y^2 = r^2$.

Page 251

1. $\cos 60° = \frac{1}{2}$, $\sin 60° = \frac{1}{2}\sqrt{3}$;
   $\cos 120° = -\frac{1}{2}$, $\sin 120° = \frac{1}{2}\sqrt{3}$;
   $\cos 240° = -\frac{1}{2}$, $\sin 240° = -\frac{1}{2}\sqrt{3}$;
   $\cos 300° = \frac{1}{2}$, $\sin 300° = -\frac{1}{2}\sqrt{3}$.
2. (i) $\frac{1}{2}$, $\frac{1}{2}$. They are equal.
   (ii) $\frac{1}{2}\sqrt{3}$, $-\frac{1}{2}\sqrt{3}$. Same size, opposite signs.
3. $1/\sqrt{2}$, $1/\sqrt{2}$.
4. $\cos 225° = -1/\sqrt{2}$, $\sin 225° = -1/\sqrt{2}$.

The Search for Pattern

Page 256

First column: c, s, $-s$, $-c$, $-c$, $-s$, s, c.
Second column: s, c, c, s, $-s$, $-c$, $-c$, $-s$.

Page 280

1. Lifts the triangle 5 units.
2. Moves the triangle 'north-east'.
3. Enlarges the triangle.
4. Produces a broader triangle with the same height.
5. Produces a taller triangle with the same breadth.
6. A reflection, as in an upright mirror, placed on the line $x = 4$.
7. A reflection, as in a horizontal mirror on the line $y = 4$.
8. See discussion following exercise.

Page 283

1. $x = X$, $y = Y + 5$.
2. $x = X + 5$, $y = Y + 5$.
3. $x = 2X$, $y = 2Y$.
4. $x = 2X$, $y = Y$.
5. $x = X$, $y = 2Y$.
6. $x = 8 - X$. $y = Y$.
7. $x = X$, $y = 8 - Y$.
8. $x = 8 - X$, $y = 8 - Y$.

Page 287

1. $y = x$. 2. $y = x$. 3. $y = x$. The result is reasonable. The transformation slides the line along itself, so the form of the equation is unaltered.
4. $y = 2x + 1$.

Page 289

1. $y = x^2 - 2x + 1$.
2. $y = x^2 - 4x + 4$.
3. $y - 1 = (x - 1)^2$ or $y = x^2 - 2x + 2$.
4. $(2, 1)$. $x = X + 2$, $y = Y + 1$, is translation required in both second and third questions.

Answers to Exercises

Page 296

$(x^2/a^2) + (y^2/b^2) = 1.$

Page 316

1. (i) Thin. Positive. Near 1. (ii) G = (1, 2).
   (iii) X −20   −10   10   20
        Y −17   −12   4   25
   (iv) 0. (v) 0. (vi) 250; 268·5; 250. (vii) 1.
   (viii) +3, −2, −6, +5. Sum, 0.
   (ix) $\sqrt{250} = 15·81$, $\sqrt{268·5} = 16·39$.
   (x) 0·96. (xi) 0·96.
2. (i) Fairly thin. Positive. Not very far from 1.
   (ii) (10, 5). (vi) 68; 21·5; 34. (vii) $\frac{1}{2}$.
   (viii) +1, −2, +3, −2. (ix) 8·24; 4·64. (x) 1·78.
   (xi) 0·89.
3. (i) Fat. Positive. It in fact resembles the egg-shaped cluster for fathers' and sons' heights, though this is not very easy to see.
   (ii) G = (4, 4). (vi) 10; 35; 10. (vii) 1.
   (viii) +5, −5, −5, +5. (ix) 3·16; 5·92.
   (x) 0·53. (xi) 0·53.
4. (i) Fat. In question 3 the y values showed a certain upward trend as x increased. That is missing here. There is no evidence of any connexion between x and y. We expect r near to 0. (ii) G = (6, 6).
   (vi) 20; 25·5; −1.
   (viii) −1/20. (viii) +4·7; −4·1; −5·9; +5·3. (ix) 4·47; 5·05.
   (x) 0·88. (xi) $r = -0·04$. No correlation.

Page 327

(i) 3. (ii) 10. (iii) 4. (iv) 6. (v) 7. (vi) 1. (vii) $\frac{1}{2}$.

*Some other Pelicans by W. W. Sawyer
are described overleaf*

Also by W. W. Sawyer

INTRODUCING MATHEMATICS I

VISION IN ELEMENTARY MATHEMATICS

Anyone who has read *Mathematician's Delight* or *Prelude to Mathematics* knows W. W. Sawyer as a mathematical lion-tamer. Figures do not merely come to life for him: they eat out of his hand.

Here he once again presents elementary mathematics in the most graphic and least terrifying way possible. As he early observes, we most of us possess a direct vision which allows us to 'see' the smaller numbers. But how to organize in our minds the chaos that lies beyond the smallest numbers is a problem that confronts the entire human race. In tackling this problem, both for those who find figures fun and, especially, for those who may be called on to teach, W. W. Sawyer offers to a wider circle methods which are already used by many good teachers – methods of visualizing, dramatizing, and analysing numbers so that the attention and understanding of children can be gained and held.

There is a boom in mathematics today. Anyone, from parent to part-time teacher, may at any moment need to understand problems in elementary arithmetic or algebra. This lively and human book can help enormously to lighten the task.

and

INTRODUCING MATHEMATICS 4
A PATH TO MODERN MATHEMATICS

Other Pelicans by W. W. Sawyer

MATHEMATICIAN'S DELIGHT
PRELUDE TO MATHEMATICS